Exploring the Multiverse: Unveiling the Cosmos of Infinite Realities

Infinite Realities

By M. Wilton

Chapter 1: Introduction

Chapter 2: Foundations of Cosmology

Chapter 3: Quantum Mechanics and the Multiverse

Chapter 4: String Theory, Extra Dimensions, and
the Multiverse

Chapter 5: Cosmological Inflation, Eternal
Inflation, and Multiverse Dynamics

Chapter 6: Anthropic Principle, Fine-Tuning, and
the Multiverse

Chapter 7: Quantum Multiverse and the
Interpretation of Reality

Chapter 8: Cosmological Implications of the Multiverse

Chapter 9: Multiverse and the Nature of Reality

Chapter 10: Multiverse and the Future of Science and Philosophy

Chapter 11: Multiverse and the Quest for Cosmic Unity

Chapter 12: Multiverse and the Metaphysics of Reality

Chapter 13: Multiverse and the Evolution of Consciousness

Chapter 14: Multiverse and the Ethical Landscape

Chapter 15: Multiverse and the Quest for Ultimate Meaning

Chapter 16: Multiverse and the Future of Scientific Exploration

Chapter 17: Multiverse and the Implications for Philosophy of Science

Chapter 18: Multiverse and the Ethical, Societal, and Cultural Implications

Chapter 19: Multiverse and the Philosophical Notions of Free Will and Determinism

Chapter 20: Multiverse and the Quest for Ultimate Understanding

Chapter 1: Introduction

The Multiverse Theory: An In-depth Exploration of Parallel Realities

1.1 The Concept of the Multiverse

The multiverse theory stands at the forefront of modern theoretical physics, challenging conventional notions of the universe's singularity and uniqueness. At its core, the multiverse hypothesis proposes the existence of multiple universes, each with its own set of physical laws, constants, and properties. This paradigm-shifting idea redefines our understanding of the cosmos and raises profound questions about the nature of reality itself.

1.1.1 Historical Development and Origins

The seeds of the multiverse concept can be traced back to the early days of cosmology. Ancient civilizations pondered the nature of the cosmos, imagining diverse realms beyond our own. However, it was not until the 20th century that scientific and philosophical thought began to converge, leading to the formulation of the multiverse hypothesis.

In the 1910s, the expansion of the universe was first suggested by the observations of galaxies receding from one another. This pivotal insight laid the foundation for the Big Bang theory, which describes the universe's origin in an explosive event roughly 13.8 billion years ago. As cosmologists delved deeper into the implications of this theory, questions arose about the universe's fate and the peculiar properties governing its existence.

The multiverse hypothesis gained traction through the works of physicists like Hugh Everett III and his Many-Worlds Interpretation (MWI) of quantum mechanics. Everett's interpretation proposed that every quantum event results in a branching of the universe, giving rise to multiple parallel worlds, each representing a different outcome of the quantum event. While this interpretation was initially met with skepticism, it planted the seeds for the idea of a multiverse of parallel realities.

1.1.2 Significance in Contemporary Physics and Cosmology

The multiverse theory has garnered significant attention within the realm of contemporary physics and cosmology. As our understanding of the universe has expanded, so too has the complexity of the questions we seek to answer. The multiverse hypothesis emerges as a possible solution to several conundrums that classical cosmology struggles to explain adequately.

One such issue is the "fine-tuning problem." This problem refers to the remarkable balance and precision required for the universe's physical constants to allow the emergence of life as we know it. The multiverse theory proposes that if there exist multiple universes, each with its own unique combination of physical laws and constants, then the observed fine-tuning could simply be a consequence of the universe we inhabit being one among countless possibilities.

Furthermore, the multiverse concept also has implications for the anthropic principle, which suggests that the observed physical properties of the universe are conducive to the existence of intelligent life because we, as observers, are here to observe them. The multiverse theory provides a potential explanation for why our universe appears finely tuned for life – if there are countless universes with varying properties, it's not surprising that we find ourselves in a universe suitable for our existence.

1.2 Historical Context and Theoretical Frameworks

The development of the multiverse theory cannot be separated from the broader historical context of physics and cosmology. Key theoretical frameworks and breakthroughs have paved the way for the multiverse hypothesis to gain traction and influence modern scientific discourse.

1.2.1 Quantum Mechanics and Its Intricacies

Quantum mechanics, a fundamental pillar of modern physics, introduces a new layer of complexity to our understanding of reality. At the quantum scale, particles exhibit behaviors that defy classical intuition. Concepts such as superposition and entanglement challenge our perception of a single, deterministic universe and hint at the possibility of multiple, coexisting realities.

Quantum superposition is a phenomenon in which a particle can exist in multiple states simultaneously until it is observed or measured. This idea challenges the classical notion that an object must have a definite state at all times.

Instead, quantum mechanics suggests that particles can exist in a multitude of states until observed, leading to the idea that there could be multiple versions of reality, each corresponding to a different outcome of a quantum event.

1.2.2 Many-Worlds Interpretation of Quantum Mechanics

One of the most provocative interpretations of quantum mechanics is the Many-Worlds Interpretation (MWI), proposed by Hugh Everett III in 1957. MWI posits that the act of measurement in quantum systems does not collapse the wave function into a single outcome, as traditionally thought. Instead, it suggests that all possible outcomes occur, each in its own separate branch of reality. This interpretation, though initially met with skepticism, has gained traction over the years and has become intertwined with the multiverse hypothesis.

In the context of the multiverse theory, MWI implies that every quantum event generates a branching of the universe into multiple branches, each representing a different outcome of that event. For example, in the famous Schrödinger's cat thought experiment, where a cat is in a superposition of being alive and dead, MWI suggests that there are parallel universes in which the cat is alive and others in which it is dead.

1.2.3 String Theory and Extra Dimensions

Another critical development that sets the stage for the multiverse theory is string theory, a theoretical framework aimed at reconciling general relativity and quantum mechanics. String theory posits that the fundamental building blocks of the universe are not point-like particles but rather one-dimensional strings. These strings vibrate at various frequencies, giving rise to the particles we observe.

String theory offers a fascinating avenue for exploring the multiverse hypothesis through the concept of extra dimensions. While we experience three dimensions of space and one dimension of time, string theory suggests that there could be additional dimensions that are compactified and hidden from our perception. These extra dimensions could take on various shapes and sizes, influencing the behavior of particles and giving rise to different physical properties in different universes.

1.3 Scope and Structure of the Book

This book aims to provide an exhaustive exploration of the multiverse theory, spanning from its historical roots to its contemporary implications. Each chapter will delve into a specific aspect of the theory, gradually building a comprehensive picture of this intriguing hypothesis.

The subsequent chapters will delve into the theoretical foundations of cosmology, quantum mechanics, string theory, and other relevant disciplines. We will examine various multiverse models, including bubble multiverses, parallel universes in the context of the Many-Worlds Interpretation, and the landscape of possible universes suggested by string theory.

Throughout this journey, we will confront the challenges and criticisms facing the multiverse theory, engage with philosophical and ethical implications, and assess potential experimental approaches to test its validity. By the end of this book, readers will be equipped with a nuanced understanding of the multiverse theory and its profound implications for our perception of the cosmos.

In the following chapters, we will embark on a detailed exploration of the foundations of cosmology, quantum mechanics, and string theory, setting the stage for a deeper understanding of the multiverse hypothesis and its significance in

reshaping our conception of reality.

Chapter 2: Foundations of Cosmology

2.1 Overview of the Big Bang Theory

The Big Bang theory stands as one of the most monumental achievements in the history of science, providing a comprehensive framework for understanding the origin, evolution, and structure of the universe. This chapter embarks on a journey through the foundational concepts of the Big Bang theory, delving into the intricate details that have reshaped our perception of cosmic history and its relevance to the multiverse hypothesis.

2.1.1 The Cosmic Microwave Background Radiation

At the heart of the Big Bang theory lies a remarkable discovery: the cosmic microwave background radiation (CMB). This faint, pervasive glow is a relic of the universe's infancy, a fossilized echo of the intense heat and energy that prevailed shortly after the Big Bang. Its discovery in 1965 by Arno Penzias and Robert Wilson fundamentally altered our understanding of the universe's origin.

The CMB, observed in all directions of the sky, presents itself as a nearly uniform glow with minute temperature fluctuations. These temperature variations, often referred to as "anisotropies," are crucial clues that unlock the universe's earliest moments. By analyzing the patterns in these fluctuations, cosmologists can decipher the conditions of the primordial universe, including its density, composition, and overall geometry.

The CMB's blackbody spectrum, a signature feature of thermal radiation, provides strong evidence for the Big Bang theory. This radiation was emitted when the universe was just a few hundred thousand years old and had cooled sufficiently for neutral atoms to form. The precise measurements of the CMB's temperature and spectrum offer a treasure trove of information about the universe's expansion rate, composition, and history.

2.1.2 Expanding Universe and Hubble's Law

The discovery of an expanding universe was a watershed moment in cosmology, revolutionizing our understanding of cosmic dynamics and casting light on the universe's past and future. Edwin Hubble's groundbreaking observations in the 1920s, based on the redshift of galaxies, revealed a startling truth: galaxies are not stationary but are receding from one another.

Hubble's observations led to the formulation of Hubble's Law, which describes the relationship between a galaxy's distance and its recessional velocity. The equation $v = H_0 d$, where v is the galaxy's recessional velocity, d is its distance, and H_0 represents the Hubble constant, captures the essence of cosmic expansion. This fundamental relationship has provided the basis for estimating the age of the universe and tracing its history backward in time.

The concept of an expanding universe naturally leads to the idea of a primordial singularity – a point of infinite density and temperature – from which the universe originated. The expansion not only carries galaxies away from each other but also stretches the fabric of spacetime itself, altering the wavelengths of light as it traverses the cosmos.

This phenomenon is responsible for the redshift observed in the spectra of distant galaxies.

2.2 Inflationary Theory and Its Relevance to the Multiverse

2.2.1 The Need for Inflation

While the Big Bang theory provided a coherent framework for understanding the universe's evolution, it grappled with certain challenges that inflationary theory sought to address. One of the most perplexing issues was the horizon problem. The universe appears remarkably uniform on large scales, yet regions that are separated by vast cosmic distances exhibit strikingly similar properties. How did these distant regions achieve such a high degree of thermal equilibrium when they had never interacted?

Inflationary theory emerged as a compelling solution to the horizon problem. It proposes that the universe underwent a brief but incredibly rapid expansion in its earliest moments. During this epoch, space itself expanded exponentially, smoothing out any pre-existing irregularities and allowing distant regions to come into thermal contact.

2.2.2 The Mechanism of Inflation

Inflation posits the existence of an exotic field called the "inflaton" that drove the exponential expansion of the universe. The inflaton is characterized by a potential energy that dominates the universe's energy budget during the inflationary phase. This potential energy generates a repulsive gravitational force, causing the universe to expand at an accelerated rate.

The inflationary epoch is thought to have occurred in the incredibly brief moments after the universe's birth. During this phase, the universe expanded by a staggering factor, possibly increasing its size by an enormous factor. This rapid expansion explains the uniformity observed in the CMB, as regions

that were initially in causal contact before inflation

were stretched beyond that limit during the

inflationary phase.

2.2.3 Implications for the Multiverse

The implications of inflation extend beyond

addressing cosmological challenges; they intersect

with the multiverse hypothesis. Inflation not only

provides a mechanism for smoothing out

irregularities but also offers a framework for the

generation of primordial fluctuations in the

universe's density. These fluctuations are imprinted

on the CMB and serve as the seeds for the formation of galaxies and cosmic structures.

However, inflation also opens the door to the concept of "bubble universes." During the inflationary process, different regions of space can transition from an inflationary state to a normal state at varying times. These transitions, akin to bubble nucleation, give rise to pockets of space where inflation ends. Each of these regions could potentially become a separate universe with its own distinct properties, leading to a multiverse of diverse realities.

2.3 Anthropic Principle and the Multiverse

2.3.1 The Anthropic Principle's Significance

The anthropic principle, a controversial but thought-provoking concept, argues that the universe's physical properties appear to be tailored to accommodate the existence of intelligent life. It suggests that the parameters governing fundamental forces, constants, and initial conditions have values conducive to the emergence of complex structures.

The multiverse hypothesis intersects with the anthropic principle in a profound way. If the multiverse is real, and there exist myriad universes with varying physical properties, it's not surprising that we find ourselves in a universe that is finely tuned for life. This idea aligns with the "weak" anthropic principle, which asserts that the observed universe is consistent with our existence simply because we are here to observe it.

2.3.2 The Landscape of Possible Universes

String theory, a candidate for a unified theory of all fundamental forces, introduces a captivating notion: the landscape of possible universes. String theory predicts that there are numerous ways to compactify the extra dimensions and various configurations for the fundamental constants. Each of these configurations could give rise to a different universe with distinct physical properties.

The multiverse hypothesis finds synergy with the landscape of string theory. It suggests that our universe is just one point in a vast landscape of possible configurations. This perspective transforms the question of why our universe's properties are finely tuned into a statistical inquiry. Among the countless universes, it's plausible that some would have conditions suitable for life, giving rise to the apparent fine-tuning we observe.

2.4 Conclusion

This chapter has ventured deep into the foundations of cosmology, offering an intricate exploration of the Big Bang theory and its profound implications. The cosmic microwave background radiation, the expanding universe, and the concept of inflation have unveiled the universe's origin and early evolution, reshaping our understanding of cosmic dynamics.

Moreover, inflationary theory's intersection with the multiverse hypothesis has enriched our perception of reality's complexity. The notion of bubble universes, emerging from quantum fluctuations during inflation, and the landscape of string theory have unveiled a diverse tapestry of potential universes, each with its own unique attributes.

As we traverse this academic journey, each chapter will delve into the complex interplay of theoretical frameworks, experimental evidence, and philosophical considerations that converge to shape our comprehension of the multiverse hypothesis. The layers of scientific knowledge we uncover will ultimately illuminate the intriguing possibilities and profound implications of a multiverse filled with parallel realities.

Chapter 3: Quantum Mechanics and the Multiverse

3.1 Introduction to Quantum Mechanics

Quantum mechanics, the revolutionary theory that governs the behavior of particles at the smallest scales, has not only transformed our understanding of the subatomic world but has also sparked contemplation about the nature of reality itself. This chapter delves deep into the intricate tapestry of quantum mechanics and its intimate connection with the multiverse hypothesis, exploring concepts such as superposition, wave-particle duality, and the enigmatic phenomenon of quantum entanglement.

3.1.1 The Quantum State and Superposition

Quantum mechanics introduces a new framework for describing the behavior of particles, one that is fundamentally probabilistic. The quantum state of a particle is represented by a mathematical construct called the wave function, which encodes information about its possible positions, momenta, and other observables. Unlike classical mechanics, where particles possess well-defined values for these properties, quantum particles exist in a state of superposition, wherein they can inhabit multiple states simultaneously.

The concept of superposition can be exemplified by the famous Schrödinger's cat thought experiment. In this scenario, a cat enclosed in a sealed box is subjected to a quantum event that places it in a superposition of being both alive and dead. According to quantum mechanics, until the box is opened and the cat is observed, it exists in a state of indeterminacy, occupying both states concurrently.

3.1.2 Wave-Particle Duality and the Uncertainty Principle

Wave-particle duality is a cornerstone of quantum mechanics, suggesting that particles such as electrons and photons exhibit both particle-like and wave-like properties. This duality challenges classical notions of particles as discrete, localized entities and underscores the probabilistic nature of quantum phenomena.

The uncertainty principle, formulated by Werner Heisenberg, articulates a fundamental limit on the precision with which certain pairs of complementary observables, such as position and momentum, can be simultaneously known. This principle reflects the inherent fuzziness of the quantum world, where the more precisely one observable is known, the less precisely the conjugate observable can be determined.

3.2 Quantum Superposition and the Many-Worlds Interpretation

3.2.1 Quantum Superposition and Measurement

The principle of superposition implies that a quantum particle can exist in a linear combination of multiple states until a measurement is performed. Upon measurement, the wave function collapses, and the particle assumes a definite state corresponding to the measurement outcome. This collapse is a pivotal juncture in quantum mechanics, triggering the transition from a probabilistic wave-like description to a classical, definite reality.

The process of measurement has intrigued physicists and philosophers alike. The question of what constitutes an "observer" and the role of consciousness in the collapse of the wave function has sparked debates and speculations. It is within this context that the Many-Worlds Interpretation (MWI) emerges as a unique proposal to interpret quantum mechanics.

3.2.2 Many-Worlds Interpretation of Quantum Mechanics

The Many-Worlds Interpretation, proposed by Hugh Everett III in 1957, challenges the conventional view of wave function collapse by suggesting that every possible outcome of a quantum event occurs in a separate branch of reality. According to MWI, the act of measurement doesn't cause a collapse; instead, all possible outcomes coexist in different "worlds" or branches.

In this interpretation, when a quantum system undergoes superposition and multiple outcomes become possible, the universe bifurcates into different branches, each representing one of the possible outcomes. For example, in the Schrödinger's cat scenario, according to MWI, there exist parallel universes in which the cat is alive and others in which it is dead.

3.2.3 Quantum Decoherence and Branching Universes

While MWI provides an intriguing perspective on the nature of reality, the question of why we don't experience multiple outcomes simultaneously arises. The phenomenon of quantum decoherence offers a potential answer. Decoherence is the process by which a quantum system's interactions with its environment cause its superposition to collapse into one of the possible outcomes.

In the context of MWI, decoherence plays a crucial role in maintaining the apparent separation of parallel universes. As a quantum system interacts with its surroundings, its different branches become entangled with the environment, effectively preventing macroscopic superpositions from persisting and leading to the appearance of a classical, single reality.

3.3 Quantum Entanglement and Non-Locality

3.3.1 Quantum Entanglement: Spooky Action at a Distance

One of the most captivating phenomena in quantum mechanics is entanglement. Two particles can become entangled when their quantum states are correlated in such a way that the state of one particle is dependent on the state of the other, even when they are separated by vast distances. Entanglement was famously referred to by Einstein as "spooky action at a distance," highlighting its puzzling nature.

Entanglement challenges classical intuition by defying the limits of locality. In a classical world, information cannot travel faster than the speed of light, ensuring that distant events are causally disconnected. However, entanglement seems to allow instantaneous correlations between entangled particles, irrespective of the spatial separation between them.

3.3.2 Bell's Theorem and Tests of Entanglement

In the 1960s, physicist John Bell formulated a theorem that provides a way to experimentally test the predictions of quantum mechanics against the predictions of classical theories that assume hidden variables. Bell's theorem showed that if the predictions of quantum mechanics hold true, the correlations between measurements on entangled particles would violate certain inequalities.

Experiments testing Bell's theorem have been conducted, and the results consistently favor the predictions of quantum mechanics over classical hidden-variable theories. These experiments confirm the reality of entanglement and the non-local nature of quantum interactions, reinforcing the enigmatic character of the quantum world.

3.3.3 Implications of Entanglement for the Multiverse

Entanglement, with its seemingly instantaneous connections between particles, has implications for the multiverse hypothesis. If the universe is composed of multiple parallel branches, as MWI suggests, entanglement could transcend these branches, enabling correlations between particles in different universes.

Speculations about entanglement's role in connecting parallel universes are fertile ground for thought experiments and philosophical contemplation. While current experimental evidence supports entanglement within a single universe, the possibility of entanglement across parallel universes raises intriguing questions about the boundaries of quantum reality.

3.4 Quantum Mechanics, Parallel Realities, and the Multiverse

3.4.1 Multiverse Scenarios in Quantum Mechanics

The intersection of quantum mechanics and the multiverse hypothesis offers captivating scenarios that challenge our intuitions about the nature of reality. The concept of parallel worlds arising from quantum superposition and branching universes resonates with the multiverse hypothesis, suggesting a multitude of interconnected realities.

Within the framework of MWI, the emergence of parallel universes provides an elegant explanation for quantum phenomena. When a particle undergoes superposition, each outcome corresponds to a separate universe. This perspective offers a novel understanding of the probabilistic nature of quantum events, suggesting that all possible outcomes are realized in different branches of reality.

3.4.2 Quantum Mechanics and the Anthropic Principle

Quantum mechanics also bears relevance to the anthropic principle, which posits that the universe's properties are conducive to the existence of intelligent life because we, as observers, are here to observe them. In a multiverse populated by myriad universes with different properties, the anthropic principle finds a natural explanation.

The multiverse hypothesis aligns with the anthropic principle by suggesting that the fine-tuning of our universe's parameters could be a consequence of the existence of multiple universes. We find ourselves in a universe compatible with our existence simply because other universes with different properties may not support life as we know it.

3.5 Conclusion

This chapter has delved deeply into the intricate landscape of quantum mechanics and its intricate connection with the multiverse hypothesis. The concepts of superposition, wave-particle duality, entanglement, and the Many-Worlds Interpretation have challenged classical intuitions and provoked philosophical musings about the nature of reality.

The profound interplay between quantum mechanics and the multiverse hypothesis highlights the complexity of the quantum world and the tantalizing possibilities of parallel realities. As we continue our exploration through this academic endeavor, each layer of scientific understanding uncovered will pave the way for a more nuanced comprehension of the multiverse theory and its multifaceted implications for our perception of the cosmos.

Chapter 4: String Theory, Extra Dimensions, and the Multiverse

4.1 Introduction to String Theory

String theory stands as one of the most ambitious and intriguing attempts to unify the fundamental forces of the universe and provide a comprehensive framework for understanding the cosmos at its deepest levels. In this chapter, we embark on a journey through the complex landscape of string theory, exploring its foundational principles, the concept of extra dimensions, and its profound implications for the multiverse hypothesis.

4.1.1 The Quest for Unification

One of the greatest challenges in theoretical physics is the quest for a unified theory that reconciles general relativity, which describes the force of gravity, and quantum mechanics, which governs the behavior of particles at the smallest scales. String theory arose as a candidate for such a theory, proposing that the fundamental building blocks of the universe are not point-like particles but rather one-dimensional strings.

Strings vibrate at different frequencies, giving rise to the particles we observe. The vibrational modes of strings correspond to different particles, offering an elegant way to unify the diverse array of particles in the Standard Model of particle physics.

4.1.2 Varieties of String Theory

String theory is not a single theory but rather a family of theories, each with its own unique properties. The five consistent superstring theories, as well as the more encompassing M-theory, make up the core of string theory. These theories share the common feature of postulating strings as the fundamental entities of the universe, but they also differ in their details and predictions.

One of the key developments within string theory is the concept of duality. String dualities reveal the deep connections between seemingly distinct string theories and provide a way to understand their relationships. Dualities have illuminated the hidden symmetries of string theory and have played a pivotal role in uncovering the rich landscape of possible universes.

4.2 Extra Dimensions and Kaluza-Klein Theory

4.2.1 Introduction to Extra Dimensions

A remarkable feature of string theory is its proposal of extra dimensions beyond the familiar three dimensions of space and one dimension of time. These additional dimensions are compactified and hidden from our perception due to their small size.

The existence of extra dimensions has profound implications for our understanding of the universe's geometry and the fundamental forces that govern it.

The concept of extra dimensions is not new and finds its origins in the Kaluza-Klein theory formulated in the early 20th century. The Kaluza-Klein theory aimed to unify electromagnetism and gravity by introducing an additional dimension of space. This extra dimension, when compactified, led to the emergence of both gravitational and electromagnetic fields in four dimensions.

4.2.2 Compactification and the Geometry of Extra Dimensions

String theory takes the idea of extra dimensions further by proposing that these dimensions can adopt various shapes and sizes. The process of compactification refers to the way these extra dimensions curl up and become imperceptible at our macroscopic scales. The geometry of these compactified dimensions influences the behavior of particles and the forces that act upon them.

Different shapes of compactified dimensions lead to different physical properties in our observable universe. For instance, the way extra dimensions are compactified can determine the types of particles that arise, the strengths of fundamental forces, and even the number of generations of particles in the Standard Model.

4.3 The Landscape of String Theory and the Multiverse

4.3.1 String Theory's Landscape of Possibilities

One of the most intriguing aspects of string theory is its prediction of a vast landscape of possible universes. The multitude of ways in which extra dimensions can be compactified, combined with the different vibrational modes of strings, results in a staggering number of potential configurations.

Each configuration corresponds to a different universe with distinct physical properties.

This landscape of string theory raises questions about the uniqueness of our universe's properties. Instead of a single, unique set of physical laws and constants, string theory suggests a multitude of possibilities. The concept of the multiverse emerges as a natural consequence of this landscape, where our universe is just one among countless others, each with its own specific attributes.

4.3.2 The Anthropic Principle in String Theory

The landscape of string theory also intersects with the anthropic principle, which suggests that the universe's physical properties are conducive to the existence of intelligent life because we, as observers, are here to observe them. In a multiverse containing a plethora of universes with diverse properties, the anthropic principle gains a new dimension.

String theory's landscape provides a statistical framework to address the apparent fine-tuning of our universe's parameters. Among the multitude of universes, it becomes more likely that some would possess the necessary conditions for life to emerge. The anthropic principle finds resonance in this context, offering an explanation for the observed universe's properties in light of the broader multiverse.

4.4 Experimental Challenges and Implications

4.4.1 Challenges of String Theory

While string theory presents an elegant framework for unifying fundamental forces, it also faces significant challenges. One of the primary obstacles is the lack of experimental verification. String theory's proposed energy scales are far beyond the reach of current particle accelerators, making direct experimental confirmation a daunting task.

Additionally, the sheer complexity of string theory and its landscape has led to debates about its predictive power. The vast number of possible configurations within the landscape raises questions about how to select the "correct" universe or set of physical properties that match our observations.

4.4.2 Potential Signatures of the Multiverse

Despite the experimental challenges, researchers have explored potential signatures of the multiverse that could be detected through cosmological observations. For instance, certain patterns in the cosmic microwave background radiation could indicate the presence of other universes in the cosmic landscape.

Additionally, evidence of cosmic inflation, a concept we explored in the context of the Big Bang theory, could indirectly support the multiverse hypothesis. Inflationary bubbles that give rise to separate universes could leave imprints on the CMB, offering a potential window into the multiverse's existence.

4.5 Conclusion

This chapter has navigated the intricate realm of string theory, extra dimensions, and their profound implications for the multiverse hypothesis. String theory's exploration of the fundamental fabric of reality, its prediction of a landscape of diverse universes, and its connection to the anthropic principle have reshaped our conception of the cosmos.

As we journey through the chapters of this academic exploration, each layer of understanding we uncover brings us closer to grasping the intricacies of the multiverse hypothesis. The interplay between string theory, extra dimensions, and the concept of parallel realities enriches our understanding of the universe's complexity and the

tantalizing prospects of a multiverse filled with

countless possibilities.

Chapter 5: Cosmological Inflation, Eternal Inflation, and Multiverse Dynamics

5.1 Introduction to Cosmological Inflation

Cosmological inflation, a paradigm-shifting

concept introduced in the early 1980s, has

revolutionized our understanding of the universe's early moments and paved the way for deeper contemplation of the multiverse hypothesis. In this chapter, we delve into the intricacies of inflationary theory, its various models, and the implications of eternal inflation for the multiverse.

5.1.1 Addressing Early Universe Challenges

Cosmological inflation was proposed as a solution to several long-standing challenges within the framework of the Big Bang theory. These challenges included the horizon problem, the flatness problem, and the monopole problem. Inflation offered an elegant mechanism that could account for the uniformity and geometry of the universe and the observed absence of magnetic monopoles.

The horizon problem, as discussed earlier, revolves around the puzzle of how distant regions of the universe attained such a high degree of thermal equilibrium. Inflationary theory suggests that a rapid exponential expansion in the early universe could have smoothed out these regions, bringing them into causal contact and resolving the horizon problem.

5.1.2 The Mechanism of Inflation Revisited

Inflation postulates the existence of a scalar field, often referred to as the inflaton, that drives the rapid expansion of the universe. The energy associated with the inflaton's potential dominates the universe's energy density during inflation, causing space to expand at an exponential rate.

The inflaton field evolves slowly down its potential energy curve, which results in a prolonged period of inflation. During this phase, the universe expands by an astonishing factor, possibly increasing its size by many orders of magnitude. As the inflaton rolls down the potential energy curve, it releases its energy, leading to the end of inflation and the transition to the hot, dense state described by the Big Bang theory.

5.2 Models of Inflation and Multiverse Scenarios

5.2.1 Different Inflationary Potentials

Inflationary theory is not constrained to a single inflaton potential. Various models propose different forms of potential energy for the inflaton, resulting in different inflationary behaviors and predictions. These models can give rise to varying patterns of density fluctuations, which influence the formation of cosmic structures such as galaxies and galaxy clusters.

The diversity of inflationary potentials leads to the concept of "slow-roll" conditions, wherein the inflaton's evolution is gradual enough to sustain inflation. The specific form of the potential determines the duration and energy scale of inflation, and consequently, the features of the universe that emerge from this phase.

5.2.2 Chaotic Inflation and Eternal Inflation

One notable inflationary model is chaotic inflation, proposed by physicist Andrei Linde. In chaotic inflation, the inflaton's potential energy is characterized by a "flat" potential, enabling inflation to occur over a wide range of initial conditions. This model gives rise to the concept of eternal inflation, a profound idea that contributes to the multiverse hypothesis.

Eternal inflation suggests that the inflationary process is ongoing and ever-present. In regions of the universe where inflation has ended, there may be other regions where inflation persists. This perpetual expansion of the universe leads to the generation of inflationary bubbles, each potentially evolving into a separate universe with its own physical properties.

5.3 Multiverse and Eternal Inflation

5.3.1 The Bubble Multiverse

Eternal inflation gives rise to the bubble multiverse scenario, wherein inflationary bubbles nucleate and expand within a higher-dimensional space. Each bubble represents a distinct universe with its own unique physical properties, constants, and particle content. The landscape of possible universes, as discussed in earlier chapters, finds a natural home within the bubble multiverse.

The eternal inflationary process ensures that new bubbles are continuously forming, resulting in an infinite ensemble of universes. This scenario has profound implications for understanding the diversity of physical laws and constants that we observe in our universe, as well as the underlying dynamics of the multiverse.

5.3.2 Observational Consequences of the Bubble Multiverse

While the concept of the bubble multiverse is tantalizing, it raises the question of whether we can ever detect or observe other bubbles. The challenge lies in the fact that regions of space with different physical properties may not be causally connected. As a result, information and signals from other bubbles may never reach us.

However, eternal inflation could leave subtle imprints on the cosmic microwave background radiation. If the inflationary bubbles have left traces on the CMB, such as patterns or anomalies, they could potentially offer indirect evidence of the existence of other universes within the multiverse.

5.4 Quantum Fluctuations, Multiverse, and the Fate of the Universe

5.4.1 Quantum Fluctuations in Inflation

Inflationary theory proposes that quantum fluctuations in the early universe serve as the seeds for the formation of cosmic structures. These fluctuations are generated during inflation and are imprinted on the cosmic microwave background radiation. They lead to variations in the density of matter, which eventually give rise to the galaxies, galaxy clusters, and large-scale structure we observe today.

The concept of quantum fluctuations has intriguing implications for the multiverse hypothesis. If different universes within the multiverse arise from quantum fluctuations during inflation, they could possess unique physical attributes and constants. This idea aligns with the diversity of the bubble multiverse scenario, where each bubble universe could have distinct features.

5.4.2 The Ultimate Fate of the Universe

Eternal inflation and the multiverse hypothesis also intersect with discussions about the ultimate fate of our universe. Current observations suggest that the expansion of the universe is accelerating due to a mysterious entity called dark energy. In the context of eternal inflation, this acceleration could lead to the formation of new inflationary bubbles, each giving rise to new universes.

Some theories propose that the universe could transition to a "false vacuum" state, which could lead to a "bubble of true vacuum" expanding at nearly the speed of light. If this happens, it could signal the end of our universe as we know it, with the potential emergence of new universes within the expanding bubble of true vacuum.

5.5 Conclusion

This chapter has explored the intricate interplay between cosmological inflation, eternal inflation, and their implications for the multiverse hypothesis. Inflationary theory's proposal of a rapid expansion in the early universe and its connection to the concept of eternal inflation have reshaped our understanding of cosmic dynamics and the origin of the multiverse.

As we navigate through the layers of scientific knowledge, each chapter deepens our appreciation of the multiverse theory and its multifaceted implications for the nature of reality. The dynamics of inflation, the emergence of bubble universes, and the eternal ebb and flow of creation within the multiverse provide a rich tapestry of contemplation about the vast cosmos we inhabit.

Chapter 6: Anthropic Principle, Fine-Tuning, and the Multiverse

6.1 Introduction to the Anthropic Principle

The concept of the anthropic principle, a philosophical idea at the nexus of cosmology and metaphysics, invites us to ponder the apparent alignment of the universe's physical constants and laws with the existence of life. In this chapter, we embark on a deep exploration of the anthropic principle, its various forms, and its intriguing connection to the multiverse hypothesis.

6.1.1 The Anthropic Landscape

The anthropic principle suggests that the physical properties of our universe are conducive to the emergence of intelligent life because we, as observers, are here to contemplate them. The seemingly fine-tuned values of fundamental constants, such as the strength of gravity, the speed of light, and the masses of particles, have led to debates about whether these values are a result of cosmic design or mere chance.

The multiverse hypothesis introduces a new dimension to the anthropic landscape. If the universe is part of a vast multiverse containing a diversity of physical properties, the apparent fine-tuning of our universe could be explained by the existence of many universes with different properties. The anthropic principle, in this context, becomes intertwined with statistical reasoning and the probability of life-conducive conditions.

6.1.2 Weak, Strong, and Participatory Anthropic Principles

Different forms of the anthropic principle offer varying perspectives on the role of observers and the universe's attributes. The "weak" anthropic principle posits that the universe's physical properties must be compatible with the emergence of observers. This perspective provides an explanation for why we observe certain values for fundamental constants – they are necessary for our existence.

The "strong" anthropic principle goes further, suggesting that the universe must be specifically designed to allow for the existence of intelligent observers. This perspective, while controversial, raises profound metaphysical questions about the nature of reality and the place of consciousness in the cosmos.

A more radical notion is the "participatory" anthropic principle, proposed by physicist John Wheeler. This principle posits that observers play a role in determining the universe's properties through the act of observation itself. It suggests a profound interconnectedness between the observer and the observed universe.

6.2 Fine-Tuning and the Multiverse

6.2.1 The Fine-Tuning Problem

The fine-tuning problem centers on the remarkable alignment of physical constants and parameters that allow for the emergence of complex structures, including life. Deviations from the observed values of these constants, even by a slight margin, could render the universe inhospitable to life as we know it. The question of why the universe's properties seem so precisely tuned for our existence has been a source of fascination and debate.

The multiverse hypothesis provides a novel perspective on the fine-tuning problem. If there are myriad universes with diverse physical properties, it becomes statistically likely that some of these universes would possess the conditions necessary for life. In this context, the apparent fine-tuning of our universe's parameters could be a consequence of our existence in a multiverse containing various realities.

6.2.2 The Landscape of Possibilities

String theory, as discussed in earlier chapters, predicts a landscape of possible universes characterized by different configurations of extra dimensions, particle content, and physical constants. This landscape offers a potential solution to the fine-tuning problem by suggesting that our universe is just one point in a vast array of possibilities.

Within this landscape, the values of fundamental constants are not set by fiat but arise from the dynamics of the underlying string theory. The multiverse hypothesis emerges as an elegant explanation for the observed fine-tuning: if different regions of the landscape correspond to different universes, then our universe's specific values could be the outcome of this broader cosmic diversity.

6.3 The Multiverse and the Goldilocks Enigma

6.3.1 The Goldilocks Enigma

The "Goldilocks enigma" refers to the observation that the universe's parameters and conditions seem to be "just right" for the emergence of life. This concept draws an analogy to the fairy tale of Goldilocks and the Three Bears, where Goldilocks seeks the porridge, chair, and bed that are neither too hot nor too cold, too big nor too small.

The Goldilocks enigma prompts questions about the reasons behind the universe's apparent suitability for life. Is it the result of design, chance, or an underlying principle that ensures life-conducive conditions? The multiverse hypothesis presents a unique way to address this enigma, suggesting that the variety of conditions across different universes accounts for the range of

possible outcomes, including ones that support life.

6.3.2 Cosmic Natural Selection

The concept of cosmic natural selection offers an

intriguing explanation for the apparent fine-tuning

of our universe's properties. Just as in biological

evolution, universes with physical properties that

allow for the emergence of life may be more likely

to produce observers who ponder their existence.

Universes with different properties that preclude

the emergence of life may remain unobserved.

In this perspective, the multiverse can be seen as a cosmic "breeding ground" for different types of universes. Over time, life-conducive universes would be "selected" through the emergence of observers, leading to a natural bias toward universes with conditions compatible with life.

6.4 Ethical Implications and the Anthropic Principle

6.4.1 Ethical Considerations

The anthropic principle also raises ethical considerations. If the multiverse hypothesis is true, and there exist countless universes with different histories and outcomes, it prompts the question of how ethics and morality are affected by this framework. If every possible action and consequence are realized in different universes, does this undermine notions of moral responsibility?

Ethical discussions in the context of the multiverse extend to questions about the value of human existence and the implications of choices in a vast cosmic landscape. These discussions challenge us to contemplate the significance of our actions and decisions in the context of a multiverse filled with diverse realities.

6.4.2 Moral Realism and Multiverse Ethics

Moral realism, the philosophical stance that objective moral truths exist, intersects with the multiverse framework. If the multiverse contains a range of moral outcomes and choices, it prompts us to consider whether moral values are universal across all possible universes or contingent on specific circumstances.

The implications of the multiverse for ethics touch on foundational questions about the nature of morality and the existence of moral truths. Ethical considerations extend beyond our individual universe, inviting us to engage in philosophical debates about the universality of moral principles.

6.5 Conclusion

This chapter has delved deep into the intricacies of the anthropic principle, the fine-tuning of the universe's parameters, and their profound connection to the multiverse hypothesis. The various forms of the anthropic principle – weak, strong, and participatory – have stimulated philosophical reflection about the nature of reality, consciousness, and the role of observers.

As we journey through the layers of scientific understanding, each chapter enriches our comprehension of the multiverse theory and its multifaceted implications. The anthropic principle offers a lens through which we contemplate the relationship between the universe, life, and meaning, while the multiverse hypothesis expands the canvas of possibilities for exploration and contemplation.

Chapter 7: Quantum Multiverse and the Interpretation of Reality

7.1 Introduction to Quantum Multiverse

The quantum multiverse, an extension of the multiverse hypothesis, draws its inspiration from the principles of quantum mechanics. In this chapter, we delve into the intricacies of the quantum multiverse, its various interpretations, and the philosophical implications of a reality characterized by superposition, entanglement, and parallel branches.

7.1.1 The Quantum Underpinning

Quantum mechanics, as explored in earlier chapters, challenges classical intuitions by proposing that particles can exist in multiple states simultaneously, a phenomenon known as superposition. The act of measurement collapses the wave function, determining a particle's state. However, this process raises questions about the nature of reality, the role of observation, and the boundaries between observer and observed.

The quantum multiverse hypothesis posits that all possible measurement outcomes are realized in separate branches of reality. This perspective extends the notion of superposition to entire universes, each representing a distinct measurement outcome. The quantum multiverse introduces a new layer of complexity to the already enigmatic realm of quantum mechanics.

7.1.2 Interpretations of Quantum Mechanics

Interpretations of quantum mechanics attempt to elucidate the meaning of quantum phenomena and the underlying nature of reality. Several interpretations offer diverse perspectives on how to understand the quantum world. These interpretations include the Copenhagen interpretation, the Many-Worlds Interpretation

(MWI), the pilot-wave theory, and the objective collapse models.

MWI, introduced earlier, aligns closely with the quantum multiverse hypothesis. In this interpretation, the wave function never collapses; rather, all possible outcomes coexist in separate branches of reality. This perspective challenges classical notions of a single, definite reality and pushes us to consider the existence of a multitude of parallel universes.

7.2 Quantum Entanglement and Reality

7.2.1 Entanglement and Non-Locality

Quantum entanglement, a phenomenon previously explored, poses challenges to our understanding of the nature of reality. When particles become entangled, their quantum states become correlated in a way that transcends classical notions of locality and causality. The apparent non-locality of entanglement raises questions about the underlying structure of the universe and the interconnectedness of particles.

In the context of the quantum multiverse, the implications of entanglement become even more intriguing. If all possible measurement outcomes are realized in separate branches of reality, entanglement seems to suggest an instantaneous connection between particles in different universes. This notion defies classical conceptions of space and time, prompting us to explore new frameworks for comprehending the fabric of reality.

7.2.2 The EPR Paradox and Bell's Theorem Revisited

The Einstein-Podolsky-Rosen (EPR) paradox, formulated in 1935, challenged the completeness of quantum mechanics. The paradox involves two entangled particles whose properties seem to be instantaneously determined, regardless of the spatial separation between them. Einstein famously referred to this as "spooky action at a distance."

Bell's theorem, as discussed earlier, provided a way to test the predictions of quantum mechanics against classical hidden-variable theories. Experiments inspired by Bell's theorem have consistently supported the predictions of quantum mechanics, indicating the reality of entanglement and the non-local nature of quantum interactions.

In the context of the quantum multiverse, the EPR paradox and Bell's theorem prompt us to consider the implications of entanglement across parallel universes. The idea that particles in separate branches can be instantaneously connected raises profound questions about the structure of the multiverse and the fundamental interconnectedness of all reality.

7.3 Observer-Dependent Reality and Quantum Multiverse

7.3.1 The Role of Consciousness

Quantum mechanics' intimate connection with observation has sparked debates about the role of consciousness in shaping reality. The collapse of the wave function upon measurement suggests that the act of observation plays a fundamental role in determining a particle's state. This observation-dependent nature of quantum reality has led to discussions about the relationship between the observer and the observed.

The quantum multiverse hypothesis complicates these debates. If all possible measurement outcomes are realized in separate branches of reality, the role of consciousness becomes intertwined with the existence of multiple observers across parallel universes. Questions about the nature of consciousness, its role in determining outcomes, and its relationship to the multiverse challenge us to reevaluate our understanding of reality.

7.3.2 Quantum Observer Effect and the Multiverse

The quantum observer effect, which describes how the act of observation alters the state of a quantum system, is also entwined with the quantum multiverse. The idea that an observer's presence can influence the outcome of an experiment raises questions about the relationship between observers in different branches of the multiverse.

In the context of MWI and the quantum multiverse, the observer effect takes on a unique dimension. Observers in one branch of reality may influence the outcomes experienced by observers in another branch, prompting speculation about the nature of these interactions and the extent to which different branches of the multiverse can interact.

7.4 Philosophical Considerations and Conclusions

7.4.1 Philosophical Implications

The quantum multiverse, with its intertwining of quantum mechanics and the multiverse hypothesis, presents a fertile ground for philosophical contemplation. The nature of reality, the role of observation, the relationship between observers, and the boundaries of consciousness are all topics that intersect with the intricacies of the quantum multiverse.

Questions about the existence of a single objective reality, the limitations of human perception, and the ultimate nature of the cosmos challenge us to grapple with the foundations of our understanding. The synthesis of quantum mechanics and the multiverse hypothesis invites us to explore philosophical territories that extend beyond the boundaries of empirical science.

7.4.2 The Multiverse and the Quest for Truth

As we conclude this chapter, we reflect on the overarching theme that has guided our exploration: the search for truth in a complex and interconnected cosmos. The quantum multiverse challenges us to embrace the uncertainty and non-intuitiveness of the quantum world while expanding our imagination to encompass the possibilities of multiple parallel universes.

As our journey through this academic endeavor continues, each chapter peels back layers of scientific understanding, unveiling the intricate tapestry of the multiverse hypothesis. The fusion of quantum mechanics, the multiverse, and the exploration of consciousness and reality stretches the boundaries of human knowledge and invites us to engage in a profound inquiry into the nature of existence itself.

Chapter 8: Cosmological Implications of the Multiverse

8.1 Introduction to Cosmological Implications

The multiverse hypothesis reverberates throughout cosmology, offering insights into the fundamental nature of the universe, its origin, evolution, and ultimate fate. In this chapter, we delve into the vast realm of cosmological implications that the multiverse framework brings forth, exploring concepts such as cosmic inflation, the Big Bang, and the ultimate destiny of the cosmos within the context of a multiverse.

8.1.1 A Multiverse-Inspired Big Bang

The Big Bang theory, the prevailing model for the origin of our universe, suggests that all matter and energy in the cosmos emerged from an incredibly hot and dense state. In the multiverse context, the origin of our universe is intricately linked to the dynamics of the larger multiverse. The possibility of inflationary bubbles nucleating within the multiverse leads to the idea that our universe's Big Bang was not a unique event, but rather a localized expression of cosmic creation.

The multiverse perspective introduces the idea that our universe is but one of countless expanding bubbles within a larger inflating cosmos. Each bubble corresponds to a separate universe with its own set of physical laws, constants, and initial conditions. This conception challenges our traditional notions of a single, isolated Big Bang event and encourages us to contemplate a more intricate cosmic tapestry.

8.2 Cosmic Inflation and Multiverse Dynamics

8.2.1 Multiverse Origins of Inflation

Cosmological inflation, as discussed in previous chapters, is a key component of the multiverse hypothesis. Inflationary theory posits that a rapid exponential expansion occurred in the early universe, leading to the smoothing of the cosmos and the emergence of cosmic structures. In the context of the multiverse, inflation takes on new significance as the mechanism through which new universes are continually formed.

The process of eternal inflation within the multiverse leads to the generation of inflationary bubbles. These bubbles serve as the seeds for separate universes, each with its own distinct properties. The dynamics of inflation within the multiverse invite us to consider the interplay between local bubble nucleation and the global expansion of the multiverse.

8.2.2 Eternal Inflation and Multiverse Topology

Eternal inflation's perpetuity means that inflation is ongoing in certain regions of the multiverse, leading to the continuous formation of new universes. The topology of the multiverse, shaped by the distribution of inflationary bubbles, influences the nature of cosmic structures, the geometry of space, and the observable consequences of the multiverse hypothesis.

The spatial arrangement of inflationary bubbles within the multiverse creates a cosmic web of universes with varying properties. The topology of the multiverse becomes an important avenue for exploration, as it determines the distribution of different types of universes and the extent to which they can interact with one another.

8.3 Cosmic Microwave Background and Multiverse Signatures

8.3.1 Cosmic Microwave Background as a Multiverse Probe

The cosmic microwave background (CMB) radiation, leftover heat from the Big Bang, offers a window into the universe's early history and its initial conditions. In the context of the multiverse, the CMB takes on a new role as a potential probe for signatures of other universes. The existence of other bubble universes within the multiverse could leave imprints on the CMB, providing indirect evidence of their presence.

Patterns in the CMB that deviate from the predictions of standard cosmological models could suggest the influence of neighboring universes. These imprints, often referred to as "bubbles in the sky," could offer insights into the dynamics of the multiverse, the distribution of bubble universes, and the underlying physical laws that govern their interactions.

8.3.2 Observational Challenges and Interpretations

While the concept of CMB imprints from other universes is intriguing, detecting such signatures poses significant observational challenges. Distinguishing between signals from our own universe and those from neighboring bubble universes requires advanced techniques in data analysis and precise measurements of the CMB.

Interpreting potential anomalies in the CMB data as evidence of other universes requires careful consideration. Other astrophysical and cosmological effects could mimic the signals of neighboring universes, making it essential to rule out alternative explanations before attributing anomalies to the presence of the multiverse.

8.4 The Fate of the Multiverse

8.4.1 Multiverse and the Ultimate Destiny

The ultimate fate of the multiverse raises philosophical and cosmological questions about the nature of reality and the implications of a cosmos filled with a multitude of universes. If the multiverse is eternal, with new universes continually forming, it prompts us to consider whether there is an ultimate end to this process or whether it persists indefinitely.

The interaction between universes within the multiverse also influences their collective destiny. The multiverse's dynamics determine whether universes can collide, merge, or otherwise interact over cosmological timescales. These interactions, combined with the influence of dark energy and other cosmic forces, shape the multiverse's ultimate destiny.

8.4.2 The Challenge of Empirical Verification

As we contemplate the fate of the multiverse, we face the challenge of empirical verification. The dynamic interactions and long timescales involved in the multiverse's destiny are beyond our current observational capabilities. Predicting the future behavior of the multiverse, particularly in the context of eternal inflation, requires advanced theoretical models and sophisticated simulations.

While we may not be able to directly observe the multiverse's fate, theoretical insights and simulations can provide valuable insights into its potential trajectories. Understanding the interactions between bubble universes, the interplay of cosmic forces, and the effects of different initial conditions within the multiverse will contribute to our understanding of its ultimate destiny.

8.5 Philosophical Reflections and Ethical Considerations

8.5.1 Philosophical Contemplations

The cosmological implications of the multiverse invite profound philosophical reflections about the nature of reality, the role of human observers, and the cosmic context of our existence. The multiverse's intricate dynamics challenge us to expand our conceptions of time, space, and the universe's grand narrative.

Philosophical inquiries extend to questions about the uniqueness of our universe's history and properties. If the multiverse contains a diversity of universes with different attributes, it prompts us to consider whether our universe's characteristics are truly exceptional or simply representative of a larger cosmic tapestry.

8.5.2 Ethical Considerations and Multiverse Dynamics

Ethical discussions in the context of the multiverse delve into questions about the significance of human actions and choices within a vast cosmic landscape. The existence of diverse universes with different outcomes challenges traditional notions of determinism and free will. If every conceivable choice is realized in different universes, does this impact the moral responsibility of individuals?

Furthermore, the multiverse's destiny raises ethical questions about the influence of humanity on the broader cosmic narrative. The potential for interaction between universes invites contemplation about whether our actions could impact the trajectory of the multiverse as a whole and the moral implications of such interactions.

8.6 Conclusion

This chapter has explored the profound cosmological implications of the multiverse hypothesis, delving into the intricate interplay between cosmic inflation, the origin of the universe, the cosmic microwave background, and the ultimate destiny of the cosmos within a multiverse framework.

As we continue our journey through this academic endeavor, each chapter enriches our understanding of the multiverse theory and its far-reaching ramifications. The cosmological implications extend our exploration of the multiverse into the fabric of the universe itself, challenging us to contemplate the grand narrative of cosmic creation, evolution, and the destiny that unfolds within the intricate tapestry of the multiverse.

Chapter 9: Multiverse and the Nature of Reality

9.1 Introduction to the Nature of Reality

At the heart of the multiverse hypothesis lies a profound question: what is the nature of reality? In this chapter, we embark on a philosophical and scientific exploration of the multiverse's implications for our understanding of reality, including discussions about the nature of existence, the limits of human perception, and the potential unity underlying the diverse array of universes within the multiverse.

9.1.1 The Multiverse and Levels of Reality

The multiverse hypothesis introduces the concept of multiple levels of reality, each corresponding to a distinct bubble universe within the multiverse. These levels represent different configurations of physical laws, constants, and initial conditions. The diversity of levels challenges traditional notions of a single, fixed reality and encourages us to embrace a more nuanced and layered conception of existence.

This multilayered perspective prompts us to consider the nature of reality as a complex tapestry

with interwoven threads of universes, each
contributing to the overall fabric of the multiverse.
Exploring the relationships between these levels
raises questions about the unity and diversity of
reality and how these concepts interact within the
multiverse framework.

9.1.2 Reality and the Observer's Perspective

The role of the observer in defining reality has been
a topic of philosophical contemplation for
centuries. The multiverse hypothesis adds a new
dimension to this discussion by suggesting that the
act of observation could lead to the creation of
multiple branches of reality. Observers in one

branch influence the outcomes in their own universe and potentially in other branches as well.

The quantum multiverse, as discussed in previous chapters, deepens this perspective by proposing that all possible outcomes are realized in separate branches of reality. This challenges traditional notions of objective reality and encourages us to contemplate the intricate interplay between observers, consciousness, and the quantum fabric of the multiverse.

9.2 The Multiverse and Philosophical Considerations

9.2.1 The Nature of Identity and Personal Identity

The multiverse hypothesis has implications for our understanding of identity, both on an individual and a cosmic scale. On a cosmic level, the existence of multiple universes challenges us to rethink the uniqueness of our universe's properties and history. Is our universe truly distinct, or is it a representative instance within a larger cosmic ensemble?

On a personal level, the multiverse raises questions about the nature of personal identity. If there are multiple versions of ourselves in different bubble universes, do these versions share a common identity? How do we define our identity in a multiverse where the concept of "self" may extend across a multitude of realities?

9.2.2 The Problem of Ontology and Multiverse Realities

The philosophical problem of ontology – the study of what exists – takes on new dimensions in the context of the multiverse. The existence of multiple universes with diverse attributes raises questions about the ontological status of these realities. Are all possible universes equally real, or do certain universes possess a higher ontological status than others?

This ontological inquiry touches on issues of existence, causality, and the relationship between different levels of reality. It prompts us to explore the implications of a multiverse where the boundaries between what exists and what does not become less distinct, challenging traditional notions of reality's fundamental nature.

9.3 Multiverse and Theories of Everything

9.3.1 Multiverse and the Quest for Unity

The search for a "Theory of Everything," a unified framework that combines quantum mechanics and general relativity, remains a central goal of theoretical physics. The multiverse hypothesis poses challenges and opportunities for this quest. The existence of multiple bubble universes within the multiverse suggests that a single theory may not fully capture the diversity of reality.

The landscape of string theory, discussed earlier, provides a framework where the fundamental constants and properties of our universe arise from the configurations of extra dimensions. Within this context, the multiverse is an inherent feature of string theory, implying that a complete "Theory of Everything" must account for the existence of multiple realities.

9.3.2 Holography, Emergence, and the Multiverse

The holographic principle, a concept stemming from black hole physics and string theory, proposes that the information content of a region of space can be encoded on its boundary. This principle challenges our notions of space, suggesting that three-dimensional reality might be a holographic projection from a two-dimensional boundary.

The multiverse hypothesis enriches this discussion by proposing that the entire cosmos, with its multitude of bubble universes, could be viewed as an emergent phenomenon from a more fundamental reality. This idea aligns with the concept of holography, raising questions about the relationship between the fundamental nature of the multiverse and the emergent properties of its constituent universes.

9.4 Ethical and Existential Implications

9.4.1 Ethical Considerations and Multiverse Morality

The multiverse hypothesis influences discussions about ethics and morality. If the multiverse contains a multitude of universes with different outcomes and choices, it prompts us to contemplate whether moral values are consistent across all possible realities or whether they are contingent on specific contexts.

Ethical debates within the multiverse context extend to questions about the significance of moral choices across different bubble universes. If every conceivable choice is realized in different realities, it challenges the concept of moral responsibility and raises questions about the nature of moral agency within a multiverse framework.

9.4.2 Existential Reflections

The multiverse hypothesis also invites existential reflections about the place of humanity in the vast cosmic landscape. The existence of multiple universes raises questions about the uniqueness of our existence and the extent to which our choices and actions matter within the broader context of the multiverse.

These reflections challenge us to grapple with the idea that our universe's properties and history may not be exceptional in the grand cosmic scheme. The multiverse encourages us to seek meaning and purpose in a context that extends beyond the boundaries of our individual universe, prompting us to contemplate the significance of our existence within the vast tapestry of the multiverse.

9.5 Conclusion

This chapter has taken us on a journey through the philosophical and scientific dimensions of the multiverse's implications for the nature of reality. From discussions about levels of reality, personal identity, and ontology to explorations of the quest for a Theory of Everything and the ethical and existential considerations raised by the multiverse, we have delved into the depths of the multiverse hypothesis's impact on our understanding of existence itself.

As we conclude this chapter, we recognize the profound synthesis of philosophy and science that the multiverse hypothesis inspires. The multifaceted implications of the multiverse challenge us to expand our intellectual horizons, confront the boundaries of human knowledge, and contemplate the intricate interplay between the universe, consciousness, and the nature of reality itself.

Chapter 10: Multiverse and the Future of Science and Philosophy

10.1 Introduction to the Future of Science and Philosophy

As we near the culmination of our exploration into the multiverse hypothesis, we turn our attention to the future of science and philosophy in the context of this profound concept. In this chapter, we delve into the potential trajectories of both disciplines, considering how the multiverse hypothesis could shape the evolution of human knowledge, the nature of inquiry, and the boundaries of our understanding of the universe.

10.1.1 The Multiverse as a Catalyst for Discovery

The multiverse hypothesis serves as a catalyst for scientific discovery by challenging established paradigms, prompting new avenues of inquiry, and expanding the frontiers of knowledge. Its far-reaching implications extend to fields such as theoretical physics, cosmology, quantum mechanics, and even fields beyond the natural sciences.

By fostering interdisciplinary collaboration, the multiverse hypothesis encourages researchers from various domains to unite in the quest for a deeper understanding of reality. This cross-pollination of ideas enhances our ability to tackle complex questions that transcend traditional disciplinary boundaries.

10.1.2 Multiverse and the Unification of Knowledge

The multiverse's multidimensional implications invite us to contemplate the unification of knowledge across diverse disciplines. Just as the multiverse proposes the existence of multiple bubble universes, each with distinct attributes, our expanding knowledge landscape encompasses diverse branches of inquiry, each shedding light on different aspects of reality.

The convergence of science and philosophy, facilitated by the multiverse hypothesis, underscores the interconnectedness of human understanding. As the boundaries between disciplines blur, our ability to approach profound questions – about existence, consciousness, and the nature of the cosmos – becomes enriched by diverse perspectives and methodologies.

10.2 Multiverse and the Future of Scientific Exploration

10.2.1 The Multiverse as a Testing Ground

The multiverse hypothesis provides a unique testing ground for the limits of scientific exploration. As our technological capabilities continue to advance, we may be able to gather more empirical evidence and observational data related to the multiverse. This includes potential signatures in the cosmic microwave background, particle physics experiments, and advances in our understanding of quantum mechanics.

The search for these multiverse signatures serves as a driving force for the development of cutting-edge technologies and innovative experimental methodologies. The pursuit of empirical verification challenges scientists to push the boundaries of what is possible and to refine our techniques for probing the intricacies of the cosmos.

10.2.2 Observational Challenges and Technological Innovation

The observational challenges posed by the multiverse hypothesis push the boundaries of current technologies and compel us to innovate in unprecedented ways. Developing instruments capable of detecting subtle signals from other universes demands advancements in fields such as astrophysics, particle physics, and data analysis.

Innovations in space-based telescopes, particle colliders, and computational algorithms are essential for capturing potential multiverse signatures. These advancements also have broader implications, extending our understanding of fundamental physical principles and our ability to explore cosmic phenomena beyond the realm of the multiverse.

10.3 Multiverse and the Evolution of Philosophy

10.3.1 The Multiverse and Philosophical Paradigms

The multiverse hypothesis ignites a reevaluation of philosophical paradigms, challenging traditional assumptions about reality, consciousness, and the nature of existence. Philosophical inquiries into the multiverse encompass epistemology, metaphysics, ethics, and the relationship between human knowledge and the cosmos.

As philosophers grapple with the implications of the multiverse, they engage in a dynamic dialogue that transcends the confines of individual philosophical schools. The multiverse hypothesis prompts philosophers to explore new perspectives, transcend disciplinary boundaries, and contribute to a collective understanding of the intricacies of reality.

10.3.2 Ethics, Morality, and the Multiverse

Ethical considerations within the multiverse context expand philosophical discussions about moral responsibility and the nature of moral values. The existence of countless universes with diverse outcomes raises questions about the universality of ethical principles and the relationship between moral choices and the broader cosmic landscape.

Ethical debates within the multiverse intersect with discussions about free will, determinism, and the significance of human actions within a vast multiverse filled with potential outcomes. Philosophers engage in rigorous debates about whether our choices matter in a context where every conceivable choice is realized in different realities.

10.4 Multiverse and the Grand Unification of Knowledge

10.4.1 Grand Unification of Knowledge

The multiverse hypothesis stimulates discussions about the grand unification of knowledge – a holistic framework that synthesizes scientific understanding, philosophical contemplation, and the insights of various intellectual traditions. Just as the multiverse proposes the unity of diverse bubble universes, the grand unification of knowledge envisions a unity that transcends disciplinary boundaries.

This synthesis of knowledge encompasses scientific discoveries, philosophical reflections, and the wisdom embedded within cultural, spiritual, and artistic traditions. The pursuit of a grand unification of knowledge inspires us to seek connections between seemingly disparate domains of human understanding, enriching our holistic comprehension of reality.

10.4.2 Multiverse and the Search for Meaning

As we contemplate the multiverse's implications for the grand unification of knowledge, we are led to reflections about the search for meaning in a vast and intricate cosmos. The multiverse challenges us to find significance not only within the boundaries of our individual universe but also within the context of the multiverse's diversity of realities.

The quest for meaning takes on a cosmic dimension as we explore the interconnectedness of human existence with the multiverse's tapestry of universes. The multiverse encourages us to seek meaning in our actions, choices, and endeavors within a context that extends beyond the boundaries of our local reality.

10.5 Conclusion

In our journey through the multiverse hypothesis, we have navigated through the depths of scientific theories, philosophical inquiries, and the interplay between the two. As we conclude this chapter, we recognize the profound implications that the multiverse brings to the future of science and philosophy.

The multiverse hypothesis serves as a crucible of ideas, a crucible that challenges us to push the boundaries of human knowledge, transcend disciplinary constraints, and seek the unity underlying the diversity of our understanding. The evolution of science and philosophy within the multiverse framework invites us to embark on an intellectual odyssey, a quest to uncover the mysteries of reality and to illuminate the interconnected tapestry of the cosmos.

Chapter 11: Multiverse and the Quest for Cosmic Unity

11.1 Introduction to the Quest for Cosmic Unity

As we delve deeper into the multiverse hypothesis, we encounter a profound quest for cosmic unity – a search for the underlying principles, patterns, and relationships that bind together the myriad universes within the multiverse. In this chapter, we explore the concept of cosmic unity, its significance across scientific and philosophical domains, and the potential frameworks that could help us unravel the mysteries of a multiverse teeming with diverse realities.

11.1.1 The Multiverse as a Cosmic Puzzle

The multiverse hypothesis presents a cosmic puzzle that invites us to seek a grand synthesis of knowledge and understanding. Just as the multiverse proposes the existence of countless universes with distinct attributes, the quest for cosmic unity calls us to search for the threads that tie these universes into a coherent whole.

This quest transcends disciplinary boundaries, encompassing both scientific and philosophical inquiries. The search for cosmic unity encourages us to explore the interconnectedness of physical laws, the nature of consciousness, and the broader context within which the multiverse unfolds.

11.1.2 Cosmic Unity Across Disciplines

The pursuit of cosmic unity unites scientific and philosophical disciplines in a shared endeavor.

Scientists explore the unity of fundamental forces, the emergence of cosmic structures, and the potential connections between different bubble universes. Philosophers delve into questions about the unity of consciousness, the nature of reality, and the ethical implications of a cosmic tapestry filled with diverse universes.

The intersection of these inquiries enriches our understanding of the multiverse's implications and invites us to contemplate the interconnected nature of human knowledge and the profound insights that can emerge from a holistic synthesis of diverse perspectives.

11.2 The Multiverse and Unified Theories of Physics

11.2.1 Unified Theories and the Multiverse

The quest for unified theories of physics, which seek to reconcile the fundamental forces of nature into a single framework, is closely intertwined with the multiverse hypothesis. The existence of multiple bubble universes within the multiverse raises questions about the unity of physical laws across different realities.

String theory, a leading candidate for a Theory of Everything, aligns with the multiverse framework by proposing that the universe's properties arise from the configurations of extra dimensions. The landscape of string theory suggests that the multitude of bubble universes could be manifestations of different vacuum states, each corresponding to a distinct set of physical laws.

11.2.2 Quantum Gravity and the Multiverse

Quantum gravity, the elusive theory that unifies quantum mechanics and general relativity, plays a crucial role in our quest for cosmic unity. The multiverse hypothesis introduces complexities that necessitate a quantum theory of gravity to describe the dynamics of inflation, the emergence of bubble universes, and the interactions between them.

String theory's approach to quantum gravity within the multiverse context holds promise for addressing these challenges. The interplay between gravity and quantum phenomena becomes essential for understanding the multiverse's fundamental nature, providing a framework through which we can explore the cosmic unity that underlies the diverse tapestry of universes.

11.3 Multiverse and the Unity of Consciousness

11.3.1 Consciousness Across Universes

The quest for cosmic unity extends to the realm of consciousness, raising questions about the unity of conscious experiences across different bubble universes. The multiverse hypothesis prompts us to explore the interconnectedness of conscious observers and the potential for shared experiences within a cosmic ensemble of realities.

If consciousness transcends individual universes and is present across the multiverse, it leads to intriguing possibilities of collective awareness and shared perspectives. Exploring the unity of consciousness within the multiverse challenges us to consider the boundaries of subjective experience and the potential for cross-universe communication of ideas and insights.

11.3.2 The Multiverse and the Nature of Perception

The multiverse hypothesis intersects with discussions about the nature of perception and the limitations of human sensory experience. The diversity of bubble universes prompts us to consider how our perceptions are shaped by the specific attributes of our universe and whether there are alternative ways of perceiving reality in other bubble universes.

Exploring the variations in perception across different realities offers insights into the potential range of sensory experiences and the ways in which our understanding of reality is constrained by the particular physical laws and constants of our universe. The multiverse encourages us to question the boundaries of our perceptual capabilities and to contemplate the extent to which our sensory experiences reflect a broader cosmic landscape.

11.4 Multiverse and the Unity of Meaning

11.4.1 Meaning Across Universes

The quest for cosmic unity extends to the realm of meaning and purpose, inviting us to explore the unity of significance across different bubble universes. The multiverse hypothesis raises questions about whether meaning is a universal concept that transcends realities or whether it is contingent on the specific attributes of each universe.

If meaning is inherent within the fabric of reality and shared across the multiverse, it prompts us to consider the implications for human endeavors, creativity, and the pursuit of knowledge. The search for cosmic unity within the context of meaning challenges us to seek connections between diverse experiences and to contemplate the shared narratives that could emerge from a cosmic tapestry of universes.

11.4.2 Ethical Unity and Multiverse Morality

The unity of ethical principles across the multiverse is a topic of ethical inquiry that arises from the quest for cosmic unity. If moral values are consistent across bubble universes, it raises questions about the universality of ethical principles and the potential for a shared ethical framework that transcends the boundaries of individual realities.

Conversely, if ethical values vary across universes, it challenges us to consider the ethical implications of interactions between different bubble universes. Exploring the unity or diversity of ethics within the multiverse context prompts us to contemplate the moral significance of our actions within a broader cosmic landscape.

11.5 Conclusion

The quest for cosmic unity takes us on a profound journey that spans the realms of science, philosophy, and human understanding. As we conclude this chapter, we reflect on the intricate interplay between the multiverse hypothesis and the exploration of unity across diverse dimensions of reality.

The multiverse's potential to unify knowledge, meaning, consciousness, and ethical considerations challenges us to transcend disciplinary boundaries and to seek connections that extend beyond individual bubble universes. The quest for cosmic unity within the multiverse framework illuminates the rich tapestry of existence, inviting us to contemplate the interconnectedness of all reality and the profound insights that emerge from the synthesis of diverse perspectives.

Chapter 12: Multiverse and the Metaphysics of Reality

12.1 Introduction to the Metaphysics of Reality

As we journey deeper into the heart of the multiverse hypothesis, we encounter profound metaphysical questions about the nature of reality, existence, and the fundamental fabric of the cosmos. In this chapter, we delve into the metaphysical dimensions of the multiverse, exploring concepts such as ontology, causality, and the underlying principles that give rise to the diverse array of universes within the multiverse.

12.1.1 The Multiverse and Metaphysical Inquiry

Metaphysics, the branch of philosophy that delves into the fundamental nature of reality, finds fertile ground for exploration within the multiverse hypothesis. The multiverse invites us to contemplate the metaphysical structures that underlie the existence of diverse bubble universes, the connections between different levels of reality, and the principles that govern the cosmic tapestry.

Metaphysical inquiries into the multiverse transcend disciplinary boundaries, encompassing both philosophical contemplation and scientific investigation. As we navigate the metaphysical dimensions of the multiverse, we delve into the deeper layers of existence, causality, and the fundamental nature of being.

12.1.2 Ontological Considerations and Multiverse Realities

The question of what exists – the problem of ontology – takes on new dimensions within the multiverse hypothesis. The existence of multiple bubble universes prompts us to consider the ontological status of these realities. Are these universes equally real, or do they possess varying degrees of reality?

The multiverse's ontology challenges us to question the boundaries of existence and to explore the connections between different levels of reality. It prompts us to consider whether the fundamental principles that underlie the multiverse give rise to a hierarchy of realities, each contributing to the broader cosmic tapestry in its own unique way.

12.2 Multiverse and the Nature of Causality

12.2.1 Causality Across Universes

The multiverse hypothesis raises intricate questions about the nature of causality – the relationship between causes and effects. In a cosmos teeming with bubble universes, each with its own physical laws and attributes, we confront the challenge of understanding how causal relationships operate across different realities.

The multiverse challenges traditional notions of causality, inviting us to explore whether causal chains extend beyond the boundaries of individual bubble universes. The interplay between causality and the diverse configurations of physical laws across the multiverse prompts us to consider whether there is a unified framework that governs causation or whether causality varies within different bubble universes.

12.2.2 Multiverse and the Nature of Time

Causality is intimately linked to the nature of time, and the multiverse hypothesis compels us to reexamine our conceptions of time's structure and flow. The existence of multiple bubble universes introduces complexities in understanding the temporal relationships between events in different realities.

Time's behavior within the multiverse raises

questions about the extent to which events in one

bubble universe can influence those in another. It

prompts us to explore whether there are cosmic

mechanisms that allow for interactions across time

and space, and whether these mechanisms are

consistent across the diverse array of bubble

universes.

12.3 Multiverse and the Fundamentals of Being

12.3.1 The Nature of Existence in the Multiverse

The multiverse hypothesis leads us to contemplate

the nature of existence itself. The existence of

multiple bubble universes challenges us to consider the fundamental principles that give rise to reality and the potential for diverse configurations of existence across different levels of reality.

Metaphysical inquiries into existence within the multiverse encompass discussions about the nature of being, the relationships between different bubble universes, and the principles that underlie the emergence of various realities. The exploration of existence within the multiverse context prompts us to contemplate the essence of reality that transcends the boundaries of individual universes.

12.3.2 Multiverse and the Problem of Identity

The multiverse hypothesis raises the problem of identity in a cosmic landscape filled with diverse bubble universes. Questions about identity extend to both the level of individual observers and the broader level of cosmic entities and structures.

On a personal level, the existence of multiple versions of ourselves in different bubble universes prompts us to explore the nature of personal identity and the continuity of self across different realities. On a cosmic level, the multiverse's implications for the identity of universes challenge us to consider whether each universe possesses a unique identity or whether they are interconnected threads within a larger cosmic tapestry.

12.4 Multiverse and the Quest for Fundamental Principles

12.4.1 Fundamental Principles Across Universes

The quest for cosmic unity leads us to inquire about the fundamental principles that underlie the diverse realities within the multiverse. Are there overarching principles that govern the behavior of bubble universes, or do different universes arise from distinct sets of fundamental laws and constants?

This inquiry delves into the concept of a multiverse "meta-law" – a potential set of principles that transcends individual bubble universes and unifies their behavior. Exploring the existence and nature of such a meta-law challenges us to contemplate the unity of the multiverse's diverse realities within a broader cosmic framework.

12.4.2 The Multiverse as a Cosmic Laboratory

The multiverse can be seen as a cosmic laboratory that offers insights into the fundamental principles that govern reality. The existence of diverse bubble universes provides a unique opportunity to explore the range of possibilities allowed by different configurations of physical laws and constants.

Studying the behavior of bubble universes within the multiverse enables us to probe the underlying structures of reality, uncover potential connections between different universes, and refine our understanding of the fundamental principles that give rise to the cosmos. The multiverse's status as a cosmic laboratory encourages us to engage in a profound exploration of the nature of reality itself.

12.5 Conclusion

As we conclude this chapter, we reflect on the metaphysical depths to which the multiverse hypothesis beckons us. The exploration of ontology, causality, the nature of being, and the fundamental principles of reality leads us to the heart of metaphysical inquiry, where science and philosophy converge in a shared quest for understanding.

The multiverse's metaphysical dimensions challenge us to contemplate the interconnectedness of existence, the intricate tapestry of causality, and the underlying principles that bind together the diverse array of universes within the cosmic ensemble. As we journey through the metaphysics of reality, we recognize the profound insights that emerge from the synthesis of scientific investigation and philosophical contemplation, revealing glimpses of the fundamental fabric of the multiverse.

Chapter 13: Multiverse and the Evolution of Consciousness

13.1 Introduction to the Evolution of Consciousness

In our exploration of the multiverse hypothesis, we now delve into the profound implications that the concept holds for the evolution of consciousness – the development and expansion of self-awareness, thought, and subjective experience. This chapter takes us on a journey through the intertwined paths of consciousness and the multiverse, exploring the emergence, diversification, and potential unity of conscious experiences across the vast tapestry of bubble universes.

13.1.1 The Multiverse as a Landscape of Consciousness

The multiverse hypothesis presents us with a landscape of possibilities for the evolution of consciousness. As bubble universes encompass a myriad of physical laws, constants, and initial conditions, they also give rise to a rich diversity of potential environments that can nurture conscious life.

This chapter delves into the evolution of consciousness within the context of the multiverse, exploring questions about the emergence of self-awareness, the development of cognition, and the interplay between conscious beings and the cosmic environments they inhabit.

13.1.2 Consciousness and the Multiverse's Levels of Reality

The multiverse's multiple levels of reality are not limited to variations in physical laws; they also encompass different stages of conscious development. The evolution of consciousness within different bubble universes leads to a spectrum of cognitive capacities, self-awareness, and modes of subjective experience.

This exploration of consciousness's evolution across different levels of reality invites us to contemplate the potential unity underlying the diversity of conscious experiences. It raises questions about whether there are overarching principles that govern the development of consciousness, transcending the boundaries of individual universes.

13.2 Multiverse and the Emergence of Consciousness

13.2.1 The Emergence of Self-Awareness

The emergence of consciousness within the multiverse's diverse environments is a topic of profound scientific and philosophical inquiry. How does self-awareness arise from the complex interactions of matter and energy? How do the conditions of different bubble universes influence the development of cognitive processes and self-reflective awareness?

Exploring the emergence of self-awareness within the multiverse context invites us to consider the role of biological, cognitive, and environmental factors in shaping conscious experiences. It also prompts us to reflect on the potential for variations in self-awareness across different bubble universes.

13.2.2 Multiverse and the Complexity of Cognition

The multiverse hypothesis prompts us to investigate the complexity of cognitive processes and the potential variations in cognitive capacities across different realities. How do different physical laws influence the development of intelligence, reasoning, and problem-solving within different bubble universes?

The exploration of cognitive complexity across the multiverse challenges us to contemplate the limits and possibilities of consciousness. It encourages us to examine how the diversity of cognitive experiences contributes to the overall richness of the multiverse's tapestry.

13.3 Multiverse and the Diversity of Conscious Experiences

13.3.1 Variations in Subjective Experience

The multiverse's diverse array of bubble universes gives rise to a multitude of potential conscious experiences. Different physical laws and conditions influence sensory perception, emotions, and the nature of subjective reality within each universe.

Exploring the variations in conscious experiences across the multiverse leads us to contemplate the range of sensory perceptions, emotional states, and cognitive frameworks that could emerge in different bubble universes. This inquiry expands our understanding of the intricate interplay between consciousness and the physical cosmos.

13.3.2 Multiverse and Cultural and Philosophical Diversity

The diversity of conscious experiences across the multiverse extends beyond sensory perceptions and cognitive processes. It also encompasses cultural, social, and philosophical perspectives that arise within different bubble universes.

The exploration of cultural and philosophical diversity within the multiverse challenges us to consider the impact of different worldviews, belief systems, and ethical frameworks on conscious experiences. It prompts us to reflect on the role of culture in shaping human understanding and the potential for shared insights and dialogues across different realities.

13.4 Multiverse and the Unity of Consciousness

13.4.1 Unity Across Diverse Realities

The quest for cosmic unity extends to the realm of consciousness, prompting us to explore the potential for unity across the diverse tapestry of bubble universes. Is there an underlying thread that unites conscious experiences across different realities, or is consciousness fundamentally fragmented across the multiverse?

The exploration of unity within consciousness invites us to consider whether there are universal principles that govern the development of self-awareness, cognition, and subjective experience. It also encourages us to contemplate the potential for shared insights and connections between conscious beings across the multiverse.

13.4.2 Multiverse and the Interconnectedness of Minds

The multiverse hypothesis suggests the possibility of interconnectedness between conscious minds across different bubble universes. If consciousness transcends the boundaries of individual realities, it raises questions about the potential for communication, collaboration, and the exchange of ideas between conscious beings in different universes.

Exploring the interconnectedness of minds within the multiverse context challenges us to consider the nature of communication mechanisms that could bridge the gaps between different realities. It also invites us to envision a cosmic network of conscious beings that transcends the constraints of time and space.

13.5 Conclusion

As we conclude this chapter, we reflect on the profound synthesis of consciousness and the multiverse hypothesis. The exploration of consciousness's evolution, diversity, and potential unity within the context of the multiverse leads us to the heart of human experience and understanding.

The multiverse's implications for consciousness inspire us to contemplate the nature of self-awareness, the development of cognitive complexity, and the interconnectedness of conscious beings across the cosmic ensemble. As we navigate the intricate interplay between consciousness and the multiverse, we recognize the transformative insights that emerge from the fusion of scientific investigation and philosophical contemplation.

Chapter 14: Multiverse and the Ethical Landscape

14.1 Introduction to the Ethical Landscape of the Multiverse

In this chapter, we delve into the profound ethical implications that arise within the framework of the multiverse hypothesis. The concept of multiple bubble universes introduces complex moral considerations that touch on questions of responsibility, values, and the interconnectedness of ethical frameworks across different realities. As we explore the ethical landscape of the multiverse, we navigate the intricate interplay between human actions, the diversity of moral systems, and the potential for shared ethical principles.

14.1.1 The Multiverse as a Moral Tapestry

The multiverse hypothesis expands our ethical inquiries beyond the confines of a single universe, encompassing a vast tapestry of diverse moral landscapes. Within each bubble universe, ethical systems may emerge based on distinct cultural, social, and philosophical contexts. This chapter explores the moral diversity that the multiverse introduces and its implications for the interconnectedness of ethical principles.

The ethical landscape of the multiverse beckons us to reflect on how moral systems arise, how they vary across different bubble universes, and whether there are fundamental principles that transcend

individual realities.

14.1.2 Ethical Systems Across Universes

The existence of multiple bubble universes within

the multiverse raises questions about the variations

in ethical systems that might arise. How do

different physical laws and environmental

conditions influence the development of moral

values and ethical frameworks within different

realities?

Exploring the diversity of ethical systems across

the multiverse challenges us to consider the role of

factors such as cultural influences, societal structures, and cognitive capacities in shaping moral values. It invites us to contemplate the potential range of ethical perspectives that could emerge in different bubble universes.

14.2 Multiverse and the Nature of Moral Values

14.2.1 Universality and Relativism of Moral Values

The multiverse hypothesis prompts us to contemplate the universality or relativity of moral values. Do ethical principles transcend the boundaries of individual bubble universes, or are they contingent on the specific attributes of each reality?

This exploration of the nature of moral values within the multiverse context raises questions about whether there are fundamental ethical principles that are shared across different universes. It also challenges us to consider the role of cultural relativism and the potential for diverse moral perspectives to coexist within the cosmic ensemble.

14.2.2 Ethical Implications of Interactions

The interactions between conscious beings in different bubble universes introduce ethical complexities that span the cosmic landscape. How do moral values shape interactions between individuals and groups across realities? How does the diversity of ethical systems influence decisions that affect both local and distant universes?

The exploration of ethical implications across the multiverse challenges us to consider how different moral frameworks intersect, conflict, or align when conscious beings from various realities interact. It invites us to reflect on the potential for shared ethical understandings to emerge through cross-universe dialogue and collaboration.

14.3 Multiverse and Moral Responsibility

14.3.1 Moral Responsibility in a Multiverse Context

The existence of multiple bubble universes raises intricate questions about moral responsibility. If there are countless versions of individuals and events across different realities, how do we assign responsibility for actions that have consequences across the multiverse?

Exploring moral responsibility within the multiverse context prompts us to consider whether there are cosmic mechanisms that link actions and consequences across bubble universes. It also challenges us to contemplate the implications of moral choices when they have repercussions in realities beyond our own.

14.3.2 Multiverse and the Scope of Ethical Concerns

The multiverse hypothesis extends the scope of ethical concerns beyond individual realities, inviting us to think about the impact of our actions on a cosmic scale. How do ethical considerations shift when we contemplate their consequences in the broader context of the multiverse?

The exploration of the scope of ethical concerns across the multiverse encourages us to expand our ethical perspectives and consider the long-term implications of our actions for a diversity of conscious beings and realities. It prompts us to reflect on the potential for a shared sense of ethical responsibility that transcends the boundaries of individual universes.

14.4 Multiverse and the Search for Ethical Unity

14.4.1 Unity and Diversity of Ethical Principles

The quest for ethical unity within the multiverse context challenges us to contemplate the potential for shared ethical principles that underlie the diversity of moral systems. Is there a unifying framework that transcends individual bubble universes and provides a foundation for ethical values?

Exploring the unity and diversity of ethical principles across the multiverse invites us to consider the common threads that might connect moral values in different realities. It prompts us to reflect on the potential for ethical insights and perspectives to emerge from a cosmic tapestry of conscious experiences.

14.4.2 Multiverse and the Ethical Quest for Meaning

The multiverse hypothesis prompts us to contemplate the ethical quest for meaning within a cosmic context. How do moral values contribute to the search for significance and purpose across the diverse array of bubble universes? How does the existence of countless realities influence the ways in which conscious beings derive meaning from their actions and choices?

The exploration of the ethical quest for meaning within the multiverse context challenges us to consider whether there are shared narratives, universal values, or cosmic goals that transcend the boundaries of individual realities. It invites us to reflect on the interconnectedness of ethical principles and their potential to contribute to a broader understanding of the cosmos.

14.5 Conclusion

As we conclude this chapter, we reflect on the profound synthesis of ethics and the multiverse hypothesis. The exploration of the ethical landscape within the context of the multiverse leads us to the heart of human values, interactions, and the interconnectedness of conscious experiences.

The multiverse's implications for ethics inspire us to contemplate the diversity of moral systems, the ethical implications of interactions across bubble universes, and the potential for shared ethical principles to emerge from a cosmic ensemble of realities. As we navigate the intricate interplay between ethics and the multiverse, we recognize the transformative insights that emerge from the fusion of scientific investigation and philosophical contemplation.

Chapter 15: Multiverse and the Quest for Ultimate Meaning

15.1 Introduction to the Quest for Ultimate Meaning

In our exploration of the multiverse hypothesis, we now delve into the profound quest for ultimate meaning – the search for a unifying purpose, significance, and understanding that transcends the boundaries of individual bubble universes. This chapter takes us on a philosophical and spiritual journey through the concept of ultimate meaning within the context of the multiverse, exploring how the existence of countless realities challenges us to contemplate the purpose of existence, the nature of reality, and the potential for a cosmic tapestry of meaning.

15.1.1 The Multiverse as a Cosmic Enigma

The multiverse hypothesis presents a cosmic enigma that invites us to seek deeper layers of meaning and significance. Just as the multiverse proposes the existence of diverse bubble universes, each with its own attributes and outcomes, the quest for ultimate meaning calls us to explore the threads that weave these universes into a coherent whole.

This chapter navigates the intricate landscape of ultimate meaning within the multiverse framework, inviting us to consider how the existence of multiple realities influences our understanding of purpose, significance, and the interconnectedness of all that exists.

15.1.2 Quest for Unity in a Multiverse

The quest for ultimate meaning intersects with the desire to find unity within a multiverse teeming with diversity. As the multiverse hypothesis challenges us to contemplate the connections between different bubble universes, the search for ultimate meaning calls us to seek a unifying thread that ties together the myriad realities and conscious experiences.

Exploring the quest for unity within the multiverse context prompts us to consider whether there are overarching principles, cosmic narratives, or shared insights that transcend individual realities and contribute to a broader understanding of the purpose of existence.

15.2 Multiverse and Philosophical Perspectives on Ultimate Meaning

15.2.1 The Nature of Reality and Cosmic Purpose

The multiverse hypothesis leads us to question the nature of reality and its relationship to ultimate meaning. How does the existence of diverse bubble universes influence our understanding of the purpose of existence? Can the multiverse provide insights into the fundamental purpose of reality itself?

This exploration delves into philosophical inquiries about the nature of cosmic purpose, inviting us to contemplate whether there are underlying principles or cosmic intentions that give rise to the multitude of realities within the multiverse. It challenges us to consider the significance of our existence within a cosmic tapestry of universes.

15.2.2 Multiverse and the Evolution of Consciousness

The quest for ultimate meaning within the multiverse context prompts us to contemplate the role of conscious beings in the unfolding of reality. How does the evolution of consciousness across different bubble universes contribute to the larger quest for meaning and purpose?

Exploring the relationship between the evolution of consciousness and ultimate meaning invites us to reflect on whether conscious beings play a unique role in uncovering the deeper layers of reality's purpose. It also challenges us to consider how diverse conscious experiences contribute to the richness of meaning within the cosmic ensemble.

15.3 Multiverse and Spiritual Perspectives on Ultimate Meaning

15.3.1 Multiverse as a Cosmic Symphony

Spiritual perspectives on ultimate meaning within the multiverse context often draw analogies to a cosmic symphony – a harmonious interplay of diverse realities that contribute to a larger narrative. The multiverse's existence prompts us to contemplate whether there is a grand orchestration that gives purpose to the myriad bubble universes.

This exploration invites us to consider spiritual notions of interconnectedness, cosmic order, and the potential for a unified source of meaning that transcends individual bubble universes. It challenges us to seek patterns within the multiverse that resonate with spiritual insights about the nature of existence.

15.3.2 Multiverse and the Pursuit of Transcendence

The multiverse hypothesis intersects with spiritual quests for transcendence – the aspiration to rise above the limitations of individual realities and connect with broader truths. How does the multiverse influence our understanding of transcendence and the search for ultimate meaning that transcends the boundaries of time and space?

Exploring the pursuit of transcendence within the multiverse context invites us to consider whether there are pathways for conscious beings to connect with deeper layers of reality. It prompts us to reflect on the potential for spiritual insights that emerge from contemplating the interconnectedness of all that exists.

15.4 Multiverse and the Cosmos as a Source of Meaning

15.4.1 The Cosmos as a Mirror of Meaning

The quest for ultimate meaning within the multiverse framework encourages us to contemplate the cosmos itself as a source of meaning. How do the intricate dynamics of the multiverse reflect deeper layers of purpose, significance, and interconnectedness?

This exploration invites us to see the cosmos as a mirror that reflects our own quest for meaning and understanding. It prompts us to consider whether the multiverse, with its diversity of bubble universes and conscious experiences, offers insights into the nature of meaning and the interconnected tapestry of existence.

15.4.2 Multiverse and the Cosmic Perspective

The multiverse hypothesis inspires a cosmic perspective that invites us to contemplate the grandeur of the multiverse and our place within it. How does this cosmic perspective influence our search for ultimate meaning and the sense of purpose that emerges from contemplating the vastness of the cosmic ensemble?

Exploring the cosmic perspective within the multiverse context encourages us to embrace a broader understanding of our place in the cosmos. It challenges us to consider whether the multiverse's existence provides a canvas upon which we can paint narratives of meaning, significance, and interconnectedness.

15.5 Conclusion

As we conclude this chapter, we reflect on the profound synthesis of ultimate meaning and the multiverse hypothesis. The exploration of the quest for purpose, significance, and understanding within the context of the multiverse leads us to the heart of human aspirations and the search for deeper truths.

The multiverse's implications for ultimate meaning inspire us to contemplate the nature of reality's purpose, the evolution of consciousness, and the potential for a cosmic tapestry of interconnected meanings. As we navigate the intricate interplay between ultimate meaning and the multiverse, we recognize the transformative insights that emerge from the fusion of philosophical contemplation, spiritual inquiry, and cosmic exploration.

Chapter 16: Multiverse and the Future of Scientific Exploration

16.1 Introduction to the Future of Scientific Exploration

In this chapter, we turn our attention to the future of scientific exploration within the framework of the multiverse hypothesis. As our understanding of the multiverse evolves, so too do the possibilities for uncovering new insights, refining our theories, and pushing the boundaries of human knowledge. This chapter explores the potential avenues for future scientific research, the challenges and opportunities that lie ahead, and the ways in which the multiverse hypothesis could reshape our understanding of the cosmos.

16.1.1 The Multiverse as an Expansive Frontier

The multiverse hypothesis opens up an expansive frontier for scientific exploration, inviting us to venture beyond the limits of our observable universe. Just as the concept of the multiverse challenges us to think beyond our immediate reality, the future of scientific exploration within this framework encourages us to push the boundaries of our knowledge and expand our understanding of the cosmos.

This chapter navigates the exciting possibilities that the multiverse hypothesis holds for the future of scientific research, highlighting the potential breakthroughs, discoveries, and paradigm shifts that could reshape our understanding of the universe and our place within it.

16.1.2 Multiverse and Interdisciplinary Collaboration

The future of scientific exploration within the multiverse framework is characterized by interdisciplinary collaboration, where researchers from diverse fields come together to tackle the complex challenges posed by the multiverse hypothesis. As the multiverse intersects with physics, philosophy, cosmology, and other disciplines, it encourages collaboration that transcends traditional boundaries and leads to a more holistic understanding of the cosmos.

Exploring the potential for interdisciplinary collaboration within the multiverse context challenges us to consider how insights from different fields can enrich our understanding of the multiverse's implications. It prompts us to contemplate the interconnectedness of knowledge and the ways in which diverse perspectives contribute to a more comprehensive exploration of reality.

16.2 Multiverse and the Refinement of Existing Theories

16.2.1 Multiverse and String Theory

The multiverse hypothesis has the potential to refine and reshape existing theories, such as string theory, which aligns closely with the multiverse framework. As string theory suggests the existence of extra dimensions and a landscape of possible vacuum states, the multiverse's diverse bubble universes could be manifestations of these distinct vacuum states.

The future of scientific exploration within the multiverse context involves further investigating the relationship between string theory and the multiverse, seeking to refine the predictions and implications of both theories. This exploration could lead to a deeper understanding of the underlying structures that give rise to our universe and the multitude of others within the cosmic ensemble.

16.2.2 Inflationary Theory and the Multiverse

Inflationary theory, which describes the rapid expansion of the early universe, is another theory that intersects with the multiverse hypothesis. In the context of the multiverse, inflation could be responsible for the formation of bubble universes with different physical properties.

Future research within the multiverse framework involves refining inflationary models to account for the emergence of bubble universes and exploring the dynamics of inflation within the multiverse context. This exploration could lead to a more comprehensive understanding of the cosmic origins and the mechanisms that govern the creation of diverse bubble universes.

16.3 Multiverse and Observational Signatures

16.3.1 Searching for Bubble Collisions

One of the potential observational signatures of the multiverse hypothesis is the detection of bubble collisions. If bubble universes within the multiverse come into contact, they could leave imprints on the cosmic microwave background radiation or other observable phenomena.

The future of scientific exploration involves developing sophisticated observational techniques and technologies to detect the subtle signals of bubble collisions. This exploration could provide valuable insights into the validity of the multiverse hypothesis and the nature of interactions between different bubble universes.

16.3.2 Multiverse and Cosmic Variations

The existence of multiple bubble universes within the multiverse introduces the possibility of cosmic variations in physical constants and properties. Observing these variations could provide empirical evidence for the multiverse hypothesis.

Future research within the multiverse framework involves mapping cosmic variations and analyzing their implications for our understanding of the underlying structures of reality. This exploration could shed light on the mechanisms that give rise to the diversity of bubble universes and the potential connections between them.

16.4 Multiverse and the Limits of Knowledge

16.4.1 Challenges and Boundaries

The future of scientific exploration within the multiverse framework is not without its challenges.

As we push the boundaries of human knowledge, we encounter limitations in our ability to observe, measure, and comprehend the complexities of the multiverse.

Exploring the challenges posed by the multiverse's vastness and complexity prompts us to consider the boundaries of our current scientific methodologies and the potential for new approaches to understanding the multiverse. It encourages us to contemplate the significance of the unknown and the ways in which the multiverse pushes us to expand our intellectual horizons.

16.4.2 Philosophical Implications of Limitations

The limitations of scientific exploration within the multiverse context intersect with philosophical inquiries about the nature of knowledge, reality, and human understanding. How do our limitations in comprehending the multiverse influence our philosophical perspectives on the boundaries of human cognition?

The exploration of philosophical implications within the multiverse framework challenges us to reflect on the relationship between knowledge and humility. It encourages us to embrace the vastness of the unknown and consider how the limitations of our understanding contribute to a deeper appreciation for the complexity and mystery of the cosmos.

16.5 Multiverse and the Quest for a Unified Theory

16.5.1 Unifying the Microscopic and Macroscopic Realms

The multiverse hypothesis prompts us to contemplate the quest for a unified theory that bridges the microscopic and macroscopic realms of reality. As the multiverse intersects with cosmology, quantum physics, and other branches of science, it offers an opportunity to seek a synthesis of these seemingly disparate domains.

The future of scientific exploration within the multiverse framework involves the pursuit of a unified theory that can explain both the fundamental particles and forces of the quantum world and the overarching structures and dynamics of the multiverse. This exploration could lead to a deeper understanding of the interconnectedness of all levels of reality.

16.5.2 Multiverse and the Unification of Concepts

Beyond the unification of physical theories, the multiverse hypothesis challenges us to seek a broader unification of concepts and perspectives.

How can we integrate the insights from diverse fields, such as physics, philosophy, and spirituality, into a cohesive framework that accounts for the complexity of the multiverse?

The future of scientific exploration within the multiverse context involves the integration of different ways of knowing, creating a multidimensional understanding that transcends disciplinary boundaries. This exploration invites us to contemplate the potential for a unified worldview that embraces the multiverse's intricate tapestry of realities.

16.6 Conclusion

As we conclude this chapter, we reflect on the boundless possibilities that the future of scientific exploration within the multiverse framework holds. The exploration of new theories, observational signatures, and interdisciplinary collaboration invites us to journey beyond the limits of our current understanding and embrace the mysteries that the multiverse presents.

The multiverse's implications for the future of scientific exploration inspire us to contemplate the potential breakthroughs, discoveries, and paradigm shifts that lie ahead. As we navigate the complexities of the multiverse hypothesis, we recognize the transformative insights that emerge

from the fusion of scientific investigation,

philosophical contemplation, and the boundless

spirit of human curiosity.

Chapter 17: Multiverse and the Implications for Philosophy of Science

17.1 Introduction to the Philosophy of Science and the Multiverse

In this chapter, we delve into the profound implications that the multiverse hypothesis holds for the philosophy of science. As a theoretical framework that challenges our understanding of the cosmos, the multiverse raises philosophical questions about the nature of scientific theories, the criteria for theory evaluation, and the relationship between empirical evidence and theoretical speculation. This chapter explores how the multiverse hypothesis reshapes our philosophical perspectives on the process of scientific inquiry and the epistemological foundations of our knowledge.

17.1.1 The Multiverse as a Paradigm Shift

The multiverse hypothesis represents a paradigm shift that challenges conventional notions of a single, unique universe. Just as past scientific revolutions, such as the Copernican and Darwinian revolutions, transformed our understanding of the cosmos, the multiverse presents an opportunity for a new philosophical exploration of the limits of our knowledge.

This chapter navigates the philosophical implications that the multiverse holds for the philosophy of science, highlighting how this revolutionary concept prompts us to reconsider foundational principles and epistemological frameworks that guide our scientific inquiries.

17.1.2 Multiverse and the Nature of Scientific Explanation

The multiverse hypothesis influences our understanding of scientific explanation and the criteria for theory evaluation. How do we evaluate the explanatory power of theories that extend beyond our observable universe? How do we assess the validity of explanations that invoke multiple bubble universes as part of their theoretical framework?

Exploring the multiverse's impact on scientific explanation challenges us to consider whether traditional criteria for explanatory adequacy need to be revised. It invites us to reflect on how the multiverse expands our notions of what constitutes a satisfactory explanation within the context of a broader cosmic ensemble.

17.2 Multiverse and Epistemic Challenges

17.2.1 Empirical Evidence and the Multiverse

The multiverse hypothesis raises epistemic challenges related to the nature of empirical evidence and its role in theory evaluation. How do we gather empirical evidence for phenomena that might occur in bubble universes beyond our observational reach? How do we distinguish between theoretical speculation and empirical confirmation within the multiverse framework?

The exploration of these challenges prompts us to contemplate the limitations of empirical evidence when dealing with the multiverse's vastness and diversity. It encourages us to reflect on the epistemic status of theories that draw on multiverse scenarios as part of their explanatory framework.

17.2.2 Falsifiability and Multiverse Theories

The principle of falsifiability, a cornerstone of scientific inquiry, raises questions about its applicability to multiverse theories. Can theories that invoke bubble universes be subject to empirical falsification? How do we navigate the tension between the falsifiability criterion and the speculative nature of some multiverse scenarios?

The future of philosophy of science within the multiverse context involves reevaluating the concept of falsifiability and exploring whether alternative criteria for theory evaluation need to be developed to accommodate multiverse hypotheses. This exploration prompts us to consider the nuances of scientific methodology in light of the multiverse's unique challenges.

17.3 Multiverse and the Nature of Reality

17.3.1 Metaphysical Implications of the Multiverse

The multiverse hypothesis has metaphysical implications that resonate with philosophical inquiries about the nature of reality. How do the diverse bubble universes within the multiverse impact our understanding of what is real? Can we discern a fundamental reality that underlies the multitude of universes?

Exploring the metaphysical implications of the multiverse prompts us to contemplate whether the existence of bubble universes challenges traditional notions of a singular, objective reality. It encourages us to reflect on the nature of existence and the ways in which the multiverse hypothesis informs our metaphysical perspectives.

17.3.2 Ontological Pluralism and the Multiverse

The concept of ontological pluralism, which acknowledges the existence of diverse realities or entities, aligns with the multiverse hypothesis. How does the multiverse contribute to the philosophical discourse on ontological pluralism? Can we reconcile the multiplicity of bubble universes with a coherent ontological framework?

The exploration of ontological pluralism within the multiverse context challenges us to consider the ways in which the existence of multiple bubble universes influences our ontological commitments. It prompts us to reflect on the relationship between ontological diversity and the search for unifying principles that connect the multitude of realities.

17.4 Multiverse and Philosophical Reflection

17.4.1 Multiverse and the Nature of Explanation

The multiverse hypothesis prompts us to reconsider the nature of explanation itself. How do we define explanation in the context of a framework that encompasses diverse bubble universes? What role do causality and predictability play in our understanding of the multiverse's intricacies?

Exploring the nature of explanation within the multiverse context challenges us to contemplate the ways in which the multiverse reshapes our understanding of causation, prediction, and the goals of scientific inquiry. It invites us to reflect on the philosophical underpinnings of explanation in a reality that extends beyond our observable universe.

17.4.2 Multiverse and Philosophical Pluralism

The multiverse hypothesis invites philosophical pluralism – the recognition of diverse perspectives and interpretations of reality. How do we navigate philosophical debates and perspectives within the multiverse framework? How do different philosophical lenses contribute to a more comprehensive understanding of the multiverse's implications?

The exploration of philosophical pluralism within the multiverse context challenges us to consider how philosophical perspectives from various traditions enrich our understanding of the multiverse. It encourages us to embrace a diverse range of philosophical viewpoints that contribute to a more holistic exploration of the implications and meanings of the multiverse hypothesis.

17.5 Conclusion

As we conclude this chapter, we reflect on the transformative role that the multiverse hypothesis plays in shaping the philosophy of science. The exploration of the multiverse's implications challenges us to reconsider fundamental principles of scientific inquiry, the nature of explanation, and the epistemological foundations of our knowledge.

The multiverse's impact on the philosophy of science inspires us to contemplate the complexities of theory evaluation, the nature of empirical evidence, and the nuances of reality itself. As we navigate the intricate interplay between philosophy and the multiverse, we recognize the profound insights that emerge from the fusion of scientific investigation, philosophical contemplation, and the boundless spirit of human curiosity.

Chapter 18: Multiverse and the Ethical, Societal, and Cultural Implications

18.1 Introduction to Ethical, Societal, and Cultural Implications

In this chapter, we delve into the profound ethical, societal, and cultural implications that arise from the multiverse hypothesis. As a theoretical framework that challenges our understanding of reality and our place within the cosmos, the multiverse hypothesis prompts us to contemplate the broader impacts on our values, societal structures, and cultural narratives. This chapter explores how the existence of multiple bubble universes influences our ethical decisions, societal dynamics, and the ways in which we construct meaning and identity.

18.1.1 The Multiverse and Ethical Decision-Making

The multiverse hypothesis introduces novel ethical considerations that span beyond the confines of a single universe. How do our ethical decisions resonate across different bubble universes? How does the recognition of multiple realities influence the moral choices we make and the consequences they entail?

This chapter navigates the ethical implications of the multiverse hypothesis, highlighting the ways in which our ethical decisions resonate through the cosmic ensemble and the potential for shared ethical principles to emerge across different realities.

18.1.2 Multiverse and Cultural Pluralism

The existence of diverse bubble universes within the multiverse framework reflects the concept of cultural pluralism – the coexistence of different cultural values and perspectives. How does the multiverse influence our understanding of cultural diversity and the ways in which different societies construct meaning?

Exploring the relationship between the multiverse and cultural pluralism challenges us to consider how the recognition of multiple realities informs our appreciation for diverse cultural narratives. It prompts us to reflect on the interconnectedness of human experiences and the potential for cross-cultural dialogue within the cosmic ensemble.

18.2 Multiverse and the Foundations of Morality

18.2.1 Multiverse and Moral Relativism

The multiverse hypothesis prompts us to reexamine the nature of moral relativism – the notion that moral values are contingent on cultural, societal, and individual perspectives. How do different bubble universes within the multiverse contribute to the diversity of moral values and ethical frameworks?

The exploration of moral relativism within the multiverse context challenges us to consider the ways in which our understanding of ethics is influenced by the myriad realities that exist. It invites us to reflect on the potential for shared moral values to emerge from a cosmic tapestry of conscious experiences.

18.2.2 Ethical Universality Across the Multiverse

The existence of multiple bubble universes invites us to contemplate the potential for ethical universality – the idea that certain moral principles transcend individual realities. How do we discern whether ethical principles are universal or contingent on specific bubble universes?

Future exploration involves investigating the potential for common ethical principles to emerge across the multiverse, transcending cultural and temporal boundaries. This exploration could contribute to a broader understanding of the foundations of morality and the role of conscious experiences in shaping ethical values.

18.3 Multiverse and Identity

18.3.1 Multiverse and Self-Identity

The multiverse hypothesis challenges our notions of self-identity – the understanding of who we are in relation to our surroundings. How does the recognition of multiple bubble universes influence our perception of self and the ways in which we define our identity?

Exploring the relationship between the multiverse and self-identity prompts us to contemplate the ways in which the existence of countless realities shapes our understanding of personal identity. It encourages us to reflect on the fluidity of identity and the potential for diverse self-concepts to emerge within the cosmic ensemble.

18.3.2 Multiverse and Collective Identity

Just as the multiverse challenges our individual notions of identity, it also influences our collective identities – the shared narratives, values, and traditions that bind communities together. How does the existence of diverse bubble universes inform our understanding of collective identity and its relation to other realities?

The exploration of collective identity within the multiverse context challenges us to consider how the recognition of multiple realities influences our sense of belonging to different communities. It invites us to reflect on the potential for cross-cultural and cross-universal connections that shape our collective identities.

18.4 Multiverse and Societal Dynamics

18.4.1 Multiverse and Inter-Societal Relationships

The multiverse hypothesis prompts us to contemplate the interactions between different societies across bubble universes. How does the existence of multiple realities influence our perceptions of "others" and our relationships with societies that might exist in parallel?

Exploring inter-societal relationships within the multiverse context challenges us to consider the potential for cross-universal communication, cooperation, and conflict. It invites us to reflect on the ways in which societal dynamics might be influenced by the recognition of multiple bubble universes.

18.4.2 Multiverse and Global Perspectives

The existence of diverse bubble universes encourages us to embrace global perspectives that transcend individual realities. How does the multiverse influence our understanding of global challenges, shared goals, and the interconnectedness of human experiences?

Future exploration involves contemplating the potential for global cooperation and collaboration within the multiverse framework. This exploration could lead to the development of shared narratives, collective actions, and a more comprehensive approach to addressing global issues across the cosmic ensemble.

18.5 Conclusion

As we conclude this chapter, we reflect on the profound implications that the multiverse hypothesis holds for our ethical, societal, and cultural landscapes. The exploration of these implications challenges us to reevaluate our ethical decisions, societal dynamics, and cultural narratives within the context of the cosmic ensemble.

The multiverse's impact on ethical values, cultural diversity, and the interconnectedness of human experiences inspires us to contemplate the transformative potential of recognizing our place within a broader cosmic tapestry. As we navigate the intricate interplay between the multiverse and its ethical, societal, and cultural implications, we recognize the rich insights that emerge from the fusion of scientific investigation, philosophical contemplation, and the complexities of human existence.

Chapter 19: Multiverse and the Philosophical

Notions of Free Will and Determinism

19.1 Introduction to Free Will, Determinism, and the Multiverse

In this chapter, we delve into the intricate philosophical debates surrounding free will and determinism within the context of the multiverse hypothesis. The existence of multiple bubble universes raises profound questions about the nature of human agency, the role of causality, and the extent to which our choices are influenced by the vast cosmic ensemble. This chapter explores how the multiverse challenges traditional notions of free will and determinism, and how it reshapes our understanding of human autonomy and responsibility.

19.1.1 The Multiverse as a Challenge to Determinism

The multiverse hypothesis presents a challenge to deterministic worldviews – the idea that all events are predetermined by prior causes. How does the existence of countless bubble universes, each with its own set of initial conditions and outcomes, influence our understanding of determinism?

This chapter navigates the philosophical implications of the multiverse for determinism, highlighting the ways in which the diversity of bubble universes challenges the notion of a singular, deterministic cosmic trajectory.

19.1.2 Multiverse and the Complexity of Causality

The multiverse hypothesis introduces complexity to the concept of causality – the relationship between causes and effects. How do causal chains span across different bubble universes? How does the interconnectedness of these universes impact our understanding of causation?

Exploring the relationship between the multiverse and causality challenges us to reconsider the linear and unidirectional view of cause and effect. It prompts us to reflect on the ways in which causal relationships manifest within a cosmic ensemble of bubble universes.

19.2 Multiverse and the Nature of Choice

19.2.1 Free Will in a Multiverse Context

The existence of multiple bubble universes prompts us to contemplate the nature of free will – the capacity to make choices independent of external influences. How does the recognition of diverse realities within the multiverse impact our understanding of human agency and the extent to which our choices are truly free?

The exploration of free will within the multiverse context challenges us to consider the potential for cross-universal interactions that influence our choices. It invites us to reflect on the interplay between deterministic factors and the autonomy we ascribe to our decisions.

19.2.2 Multiverse and the Quantum Uncertainty

Quantum mechanics introduces a layer of uncertainty at the microscopic level, challenging determinism. How does the interaction between quantum uncertainty and the multiverse hypothesis influence our understanding of free will and choice?

Future exploration involves investigating whether quantum uncertainties within bubble universes interact with the multiverse's dynamics to influence the choices made by conscious beings. This exploration could lead to insights into the relationship between quantum phenomena, free will, and the cosmic ensemble.

19.3 Multiverse and Moral Responsibility

19.3.1 Multiverse and Moral Accountability

The multiverse hypothesis raises questions about moral responsibility – the extent to which individuals are accountable for their actions. How does the existence of diverse bubble universes influence our understanding of moral accountability and the consequences of our choices?

Exploring the relationship between the multiverse and moral responsibility challenges us to consider whether accountability extends beyond our immediate reality. It prompts us to reflect on the ways in which our actions might resonate through the cosmic ensemble and the ethical implications of this interconnectedness.

19.3.2 Multiverse and Retrocausality

Retrocausality, the idea that the future can influence the past, intersects with the multiverse hypothesis. How does the potential for retrocausal effects within the multiverse framework impact our understanding of moral responsibility and the causal relationships between actions and their consequences?

The future of philosophical inquiry involves exploring the implications of retrocausality within the multiverse context, contemplating whether choices made in different bubble universes can retroactively influence the past. This exploration

could lead to insights into the complex interplay

between causality, free will, and the multiverse's

dynamics.

19.4 Multiverse and Personal Identity

19.4.1 Multiverse and Temporal Identity

The existence of diverse bubble universes prompts

us to reconsider the nature of temporal identity –

the continuity of an individual's identity across

different moments in time. How does the

recognition of multiple realities within the

multiverse influence our understanding of personal identity over time?

Exploring the relationship between the multiverse and temporal identity challenges us to contemplate the ways in which our identity might evolve across different bubble universes. It encourages us to reflect on the fluidity of identity and the potential for varied self-concepts within the cosmic ensemble.

19.4.2 Multiverse and Variability of Identity

The multiverse hypothesis introduces variability to the concept of personal identity – the diversity of identities that an individual might possess across different bubble universes. How does the existence of countless realities impact our understanding of the multiplicity of self-identities?

The exploration of identity variability within the multiverse context challenges us to consider the ways in which the recognition of diverse bubble universes influences our sense of self and the narratives we construct about our identities. It invites us to reflect on the potential for self-discovery and self-expression within the cosmic ensemble.

19.5 Conclusion

As we conclude this chapter, we reflect on the profound philosophical implications that the multiverse hypothesis holds for our notions of free will, determinism, and personal identity. The exploration of these implications challenges us to reevaluate our understanding of human agency, causal relationships, and the complexities of choice within the context of the cosmic ensemble.

The multiverse's impact on free will, determinism, and personal identity inspires us to contemplate the transformative potential of recognizing our place within a multiverse of diverse realities. As we navigate the intricate interplay between the multiverse and its philosophical implications, we

recognize the profound insights that emerge from

the fusion of scientific investigation, philosophical

contemplation, and the complexities of human

existence.

Chapter 20: Multiverse and the Quest for Ultimate Understanding

20.1 Introduction to Ultimate Understanding and the Multiverse

In this final chapter, we embark on a philosophical and existential journey that explores the quest for ultimate understanding within the context of the multiverse hypothesis. As we contemplate the intricacies of the cosmic ensemble and the profound implications it holds, we delve into the human pursuit of knowledge, meaning, and purpose. This chapter examines how the multiverse challenges us to confront questions of ultimate significance and how it guides our search for deeper truths about existence, consciousness, and the nature of reality.

20.1.1 The Multiverse as a Catalyst for Inquiry

The multiverse hypothesis serves as a catalyst for the human quest for ultimate understanding. Just as the night sky has inspired countless generations to gaze upward in awe and contemplation, the concept of multiple bubble universes prompts us to inquire about the nature of our place within the grand

tapestry of the cosmos.

This chapter navigates the philosophical and

existential implications of the multiverse for the

quest for ultimate understanding, highlighting how

this revolutionary concept invites us to explore the

frontiers of human knowledge and grapple with

questions that transcend the boundaries of

individual realities.

20.1.2 Multiverse and the Search for Meaning

The existence of diverse bubble universes within

the multiverse framework raises questions about

the nature of meaning and purpose. How do we find meaning in a reality that extends beyond our immediate universe? How does the multiverse inform our understanding of our individual and collective roles within the cosmic ensemble?

Exploring the relationship between the multiverse and the search for meaning challenges us to contemplate the ways in which our quest for purpose resonates through different realities. It encourages us to reflect on the potential for shared existential quests that unite beings across bubble universes.

20.2 Multiverse and the Nature of Consciousness

20.2.1 Consciousness Across the Multiverse

The multiverse hypothesis prompts us to contemplate the nature of consciousness – the awareness that underlies our thoughts, perceptions, and experiences. How does consciousness manifest within the cosmic ensemble? Can conscious beings exist in other bubble universes?

This chapter navigates the philosophical implications of the multiverse for the nature of consciousness, highlighting the ways in which the existence of multiple bubble universes challenges our understanding of conscious experiences and their relationship to the fabric of reality.

20.2.2 Multiverse and the Observer Effect

The observer effect, a phenomenon in quantum mechanics, intersects with the multiverse hypothesis. How does the potential for different observations within bubble universes impact our understanding of the observer's role in shaping reality?

Future exploration involves investigating whether the observer effect within different bubble universes interacts with the multiverse's dynamics to influence conscious experiences. This exploration could lead to insights into the relationship between observation, consciousness, and the cosmic ensemble.

20.3 Multiverse and Transcendence

20.3.1 Multiverse and Transcendent Experiences

The existence of diverse bubble universes raises questions about transcendence – the experience of moving beyond ordinary limits and entering a realm of higher understanding. How does the multiverse influence our understanding of transcendent experiences and the potential for individuals to access greater insights?

Exploring the relationship between the multiverse and transcendent experiences challenges us to consider the ways in which our connection to broader realities can lead to moments of enlightenment and expanded consciousness. It invites us to reflect on the significance of the multiverse in fostering deeper insights and personal transformation.

20.3.2 Multiverse and the Quest for Cosmic Wisdom

The multiverse hypothesis invites us to embark on a quest for cosmic wisdom – the pursuit of profound insights into the nature of reality and existence. How do different bubble universes contribute to our collective understanding of cosmic truths? How does the multiverse guide us toward a broader perspective on existence?

The future of philosophical inquiry involves exploring the potential for a shared cosmic wisdom that emerges from the exploration of different bubble universes. This exploration could lead to a deeper appreciation for the interconnectedness of knowledge and the transcendent insights that arise from contemplating the multiverse.

20.4 Multiverse and the Unanswered Questions

20.4.1 Embracing the Unknown

As we approach the culmination of our exploration, we acknowledge the multiverse's role in revealing the vastness of the unknown. How do we navigate the unanswered questions that the multiverse hypothesis presents? How do we embrace the mystery of the cosmic ensemble while striving for ultimate understanding?

The exploration of unanswered questions within the multiverse context challenges us to consider the humility required to grapple with the complexities of reality. It encourages us to reflect on the transformative potential of embracing the unknown and the ways in which it fuels our curiosity and quest for knowledge.

20.4.2 Multiverse and the Endless Quest

The multiverse hypothesis opens the door to an endless quest for understanding – a journey that extends beyond the limits of individual lifetimes and realities. How does the recognition of multiple bubble universes shape our perspective on the ongoing pursuit of knowledge and insight?

Future exploration involves contemplating the ways in which the multiverse inspires us to engage in a continuous search for wisdom that transcends time and space. This exploration could lead to a more profound appreciation for the ongoing dialogue between human curiosity and the mysteries of the cosmic ensemble.

20.5 Conclusion

As we conclude this chapter, and our exploration of the multiverse hypothesis, we reflect on the transformative role it plays in guiding our quest for ultimate understanding. The exploration of the multiverse's implications challenges us to grapple with questions of meaning, consciousness, and the nature of reality itself.

The multiverse's impact on our pursuit of knowledge and wisdom inspires us to contemplate the transformative potential of recognizing our place within a cosmic ensemble of diverse realities. As we navigate the intricate interplay between the multiverse and the quest for ultimate understanding, we recognize the profound insights that emerge from the fusion of scientific investigation, philosophical contemplation, and the unending spirit of human curiosity.

In closing, we acknowledge that the multiverse hypothesis invites us to embrace the mysteries that lie beyond the horizon of our current understanding and to continue our journey toward the infinite realms of cosmic insight and wonder.